MONOGRAPH NO. 16 / 1976

Conformal Projections
Based on Elliptic Functions

by L. P. LEE
formerly Chief Computer,
Department of Lands and Survey,
Wellington, New Zealand

Supplement No. 1 to *Canadian Cartographer*, Vol. 13, 1976

CARTOGRAPHICA

Published by B. V. Gutsell, Department of Geography
York University, Toronto, Canada

Printed in Canada
BY UNIVERSITY OF TORONTO PRESS

ISBN 0-919870-16-3
ISSN 0317-7173

Printed in Canada by
UNIVERSITY OF TORONTO PRESS

*This series of monographs on cartography is published
with the assistance of grants from The Social Science
Research Council of Canada, The Canada Council, The
National Research Council of Canada, The National
Commission for Cartography, and the Department of
Energy, Mines and Resources.*

Foreword

   L. P. Lee, who retired in 1974 as Chief Computer in the New Zealand
Department of Lands and Survey, is a noted authority on map projections.
His work in map projections has had three main phases.   The first cul-
minated in 1944 with his publication of a classification of map projections (along with rules for nomenclature) that has since become widely
accepted.   The second phase was a critical study of the transverse Mercator projection -- or rather projections, since Lee reminded the cartographic world that a projection name which is well-defined for the sphere
may become ambiguous when applied to the spheroid.   This phase was marked
by a number of papers published between 1945 and 1963.   The first paper
of the third phase appeared in 1965; this phase has been an exploration
of the applications of elliptic functions to conformal map projections.
The present monograph may be regarded as the culmination of this third
phase.   But it is more than that: it is the culmination of the second
phase as well, since the natural way to discuss the transverse Mercator
projections is to use elliptic functions.   Furthermore, the second and
third chapters (before elliptic functions are introduced) are a concise
introduction to the theory of conformal projections of the sphere and of
the spheroid.   The author refers to tables of elliptic functions and of
map coordinates, however, only the mathematical tables are included in
this monograph.

THOMAS WRAY
Surveys & Mapping Branch / Ottawa

An Acknowledgement

I would like the readers of this monograph to know that Mr Lee prepared his manuscript as camera copy. He typed the whole work himself, drafted the figures that illustrate it twice as the original set was lost in transit, and he assembled each page with meticulous care. The work as presented here is thus totally the author's achievement. It is a privilege having this opportunity to publish Mr Lee's work.

BERNARD V. GUTSELL
York University

Contents

I
Introduction

1. Scope of study. Most of the projections described in this study stem from the work of H. A. Schwarz who in 1864 (published 1869) showed that the interior of a circle can be conformally represented by the interior of a regular polygon. In the same paper, listed in the bibliography as Schwarz 1869a*, he gave the formula for the conformal representation of the interior of an ellipse by the interior of a circle, described in more detail in Schwarz 1869c. In 1872 he showed that the spherical triangles associated with the planes of symmetry of the regular polyhedra can be conformally represented on the infinite plane, and from this he gave formulae for the conformal representation of the sphere upon each of the five regular polyhedra. The first use of any of these projections in terrestrial cartography was by Peirce 1879, who mapped a hemisphere conformally within a square, and the transverse aspect of the same projection was described by Guyou 1887.

Schwarz expressed his projections by means of integrals, and only in the ellipse-to-circle transformation did he explicitly introduce elliptic functions. In 1925 Oscar S. Adams published *Elliptic Functions Applied to Conformal World Maps*, in which he noted that Schwarz's integral of 1864, for the case $n = 3$, can be identified with Dixon elliptic functions, and the first part of his publication is a mathematical study of these functions. The second part contains examples of conformal projections based on Dixon elliptic functions as well as some not explicitly defined in terms of elliptic functions. The projection of Peirce and Guyou was included, but Adams followed Guyou's ingenious but laborious analysis, and did not take advantage of the great simplification that results from the use of Jacobian elliptic functions. In later years, a few further examples of projections of the same general class were added by Adams 1929, 1936, Cox 1935, Lee 1965.

Adams worked with logarithms, and his methods of computation were often extremely lengthy. He made no mention of any independent checks, and his tables of coordinates contain a number of errors. Most of these are of little consequence, but there are two which visibly distort his diagrams. In many cases, Adams merely stated a formula for a projection without providing any clue as to why that formula should be chosen.

The present study is a resurvey of conformal projections based on elliptic functions, with formulae suitable for computation by desk calculator. In most cases, formulae for direct and for inverse computation are given, so that an independent check is possible. Series in powers of a complex variable have been used extensively, as they are simple and practical, and eminently suited to modern computing equipment. The study includes all the projections previously mentioned and several new examples.

New tables of Dixon elliptic functions are given, as the table provided by Adams was found inadequate. Coordinates have been computed to six decimals, finally rounded off to four decimals. In the course of this work, Adams's tables of coordinates have been recomputed.

*A personal name followed by a year is either a reference to a publication listed in the bibliography, with a letter of the alphabet when there is more than one publication in the same year (e.g. Adams 1925a), or, when no publication is listed, is an indication of the year when the work was done by the person named (e.g. Thompson 1945).

I acknowledge constructive criticism of the preliminary draft of this monograph made by Thomas Wray, whose suggestions have resulted in improvements in several of the sections.

2. Notation. The projections described in this paper are defined for a sphere of unit radius. For the construction of any such projection, the coordinates computed from the formulae herein would be multiplied by the radius of the sphere reduced to the nominal scale of the map.

The principal symbols used are the following.

ϕ geodetic or geographic latitude
ψ isometric latitude, defined by
        $d\psi = \sec\phi\, d\phi$ for the sphere,
        $d\psi = (\rho/\nu)\sec\phi\, d\phi$ for the spheroid
ϑ colatitude, $\tfrac{1}{2}\pi - \phi$
λ longitude, usually reckoned from the meridian of
        the origin of the projection
$\zeta = \psi + i\lambda$
$r = p + iq$
$w = u + iv$ } plane rectangular coordinates
$z = x + iy$
ξ coordinates of projection on infinite plane
m scale coefficient or magnification
γ convergence or rotation

In these coordinate systems, the positive real axis is directed northward, the positive imaginary axis is directed eastward, and angles are reckoned eastward (clockwise) from north, as is usual in geodetic and cartographic applications. Readers who prefer the mathematical convention with the positive real axis directed eastward and with angles reckoned counterclockwise from east may rotate and reverse the diagrams; the mathematics is not affected.

lam ϕ denotes the lambertian of ϕ, customarily called the inverse gudermannian in English works and denoted by $\mathrm{gd}^{-1}\phi$ following Cayley's use of $\mathrm{gd}\, u$ to denote the gudermannian; the latter is here denoted by lam ^{-1}u.

sm, cm, denote the Dixon elliptic functions of which K is one-third of the real period. ω and ω^2 are the complex cube roots of 1, i.e.

$$\omega = -\tfrac{1}{2}(1 - i\sqrt{3}), \qquad \omega^2 = -\tfrac{1}{2}(1 + i\sqrt{3}). \qquad (2.1)$$

In Sec. 27 only, ω is the fifth root of 1.

sn, cn, dn, denote the Jacobian elliptic functions of which k, k', K, K', are the associated moduli and quarter-periods. In the separation of the real and the imaginary parts of a Jacobian elliptic function of a complex variable $u + iv$, functions of u have the modulus k, functions of v have the complementary modulus k'. We use the abbreviations, sn $(u, k) = \mathrm{sn}\, u$, sn $(v, k') = \mathrm{sn}'v$, and similarly for the other functions. Glaisher's notation is used for reciprocals and quotients; sc $u = \mathrm{sn}\, u/\mathrm{cn}\, u$ (sometimes written tn u), sd $u = \mathrm{sn}\, u/\mathrm{dn}\, u$. en u is the elliptic integral of the second kind, usually denoted in textbooks by $E(u)$ or $E(\mathrm{am}\, u)$. E and E' are the complete elliptic integrals of the second kind. ed u is the integral of $k'^2\mathrm{nd}^2 u$.

Appendix II lists the principal formulae used in this study.

3. Schwarz's integral. In 1864 H. A. Schwarz showed that the interior of the unit circle z can be conformally represented by the interior of a regular polygon w of n sides by means of the integral

$$w = \int_0^z (1 - z^n)^{-2/n} \, dz. \tag{3.1}$$

The circumradius of a polygonal face is found by putting the upper limit of the integral equal to 1. In this definite integral, put

$$z = x^{1/n}, \qquad dz = \frac{1}{n} x^{(1-n)/n} \, dx. \tag{3.2}$$

The circumradius can then be expressed as

$$\frac{1}{n} \int_0^1 x^{(1-n)/n} (1 - x)^{-2/n} \, dx = \Gamma\left(\frac{1}{n}\right) \Gamma\left(\frac{n-2}{n}\right) \Big/ n\, \Gamma\left(\frac{n-1}{n}\right). \tag{3.3}$$

For the case $n = 3$, the transformation can be expressed in terms of Dixon elliptic functions for $\alpha = 0$, and the circumradius is equal to K, one-third of the real period of the Dixon functions. That is, the interior of the unit circle z can be conformally represented within an equilateral triangle w by

$$\mathrm{sm}\, w = z, \tag{3.4}$$

with $K = [\Gamma(\frac{1}{3})]^2/3\Gamma(\frac{2}{3}) = [\Gamma(\frac{1}{3})]^3/2\pi\sqrt{3} = 1 \cdot 766\,639$.

For the case $n = 4$, the transformation can be expressed in terms of Jacobian elliptic functions for modulus $1/\sqrt{2} = \sin 45°$, the circumradius is $K/\sqrt{2}$, and the side of the square is K, where K is the real quarter-period of the functions. That is, the interior of the unit circle z can be conformally represented within a square w by

$$\sqrt{2}\, \mathrm{sd}\, (\sqrt{2} w, \, 1/\sqrt{2}) = z, \tag{3.5}$$

with circumradius $K/\sqrt{2} = \sqrt{\pi}\Gamma(\frac{1}{4})/4\Gamma(\frac{3}{4}) = [\Gamma(\frac{1}{4})]^2/4\sqrt{2\pi} = 1 \cdot 311\,029$.

To apply these transformations to map projections, we have first to represent the sphere, or part of the sphere, within the unit circle. The next few sections therefore consider conformal representation of the sphere in general, and conformal representations within the unit circle.

We can note that the boundary of a hemisphere centred on a point of latitude ϕ_0, with longitude reckoned from the meridian of this point, is given by

$$- \cos \lambda = \tan \phi_0 \tan \phi, \tag{3.6}$$

an equation derived from the quadrantal spherical triangle whose vertices are the origin, the pole, and the boundary point (ϕ, λ).

4. Isometric coordinates on the sphere. If ϕ is the latitude and λ the longitude of a point on a sphere of unit radius, an element of arc ds on the surface at that point can be resolved into components $d\phi$ along the meridian and $\cos\phi\,d\lambda$ along the parallel, so that

$$ds^2 = d\phi^2 + \cos^2\phi\,d\lambda^2. \tag{4.1}$$

If we put
$$d\psi = d\phi/\cos\phi, \tag{4.2}$$

we then have
$$ds^2 = \cos^2\phi\,(d\psi^2 + d\lambda^2). \tag{4.3}$$

Since there is no term involving the product $d\psi\,d\lambda$, the curves of constant ψ and constant λ are orthotomic, and, when the increments $d\psi$ and $d\lambda$ are equal, they are the sides of an infinitesimal square. The set of curves therefore divides the surface into a network of infinitesimal squares, and is called an isometric system.

In this particular example of an isometric system, the curves are parallels and meridians, and ψ, a function of ϕ, is called the isometric latitude. By integration of (4.2) we have

$$\psi = \int_0^\phi \sec\phi\,d\phi = \operatorname{lam}\phi = \ln\cot\tfrac{1}{2}c, \tag{4.4}$$

where c is the polar distance or colatitude. The constant of integration is made zero by stipulating that $\psi = 0$ when $\phi = 0$. When $\phi = 90°$, then $\psi = \infty$.

Further relations follow from (4.4), e.g.,

$$\left.\begin{array}{cc}
\sinh\psi = \tan\phi, & \cosh\psi = \sec\phi, \\
\tanh\psi = \sin\phi, & \tanh\tfrac{1}{2}\psi = \tan\tfrac{1}{2}\phi, \\
\exp\psi = \sec\phi + \tan\phi = \cot\tfrac{1}{2}c, & \\
\exp(-\psi) = \sec\phi - \tan\phi = \tan\tfrac{1}{2}c, & \\
\exp\tfrac{1}{2}\psi = \sqrt{\cot\tfrac{1}{2}c}, & \exp(-\tfrac{1}{2}\psi) = \sqrt{\tan\tfrac{1}{2}c}, \\
\sinh\tfrac{1}{2}\psi = \tfrac{1}{2}(\sqrt{\cot\tfrac{1}{2}c} - \sqrt{\tan\tfrac{1}{2}c}), & \\
\cosh\tfrac{1}{2}\psi = \tfrac{1}{2}(\sqrt{\cot\tfrac{1}{2}c} + \sqrt{\tan\tfrac{1}{2}c}). &
\end{array}\right\} \tag{4.5}$$

Numerical values of some of these functions are given in Table 1.

5. Conditions for conformal representation of the sphere. The conditions for conformal representation of the sphere and of the spheroid were first explicitly set forth by Lambert 1772, and their connection with the complex variable was noted by Lagrange, as acknowledged in Lambert's paper. A more general theory was later given by Gauss. The following approach is influenced most by the work of Hotine 1946.

A conformal representation is one which preserves the shape of every infinitesimal area, so that, in the conformal representation of the sphere upon a plane, every infinitesimal square on the plane is the representation of an infinitesimal square on the sphere. Therefore some functional relation must exist between a set of isometric coordinates on the plane and a set of isometric coordinates on the sphere, and we can suppose that such a relation exists between (x, y) on the plane and (ψ, λ) on the sphere, whence

$$dx = \frac{\partial x}{\partial \psi} d\psi + \frac{\partial x}{\partial \lambda} d\lambda, \qquad dy = \frac{\partial y}{\partial \psi} d\psi + \frac{\partial y}{\partial \lambda} d\lambda. \qquad (5.1)$$

If ds is an element of arc on the sphere at azimuth A, its components in meridian and parallel respectively are

$$ds \cos A = \cos \phi \, d\psi, \qquad ds \sin A = \cos \phi \, d\lambda, \qquad (5.2)$$

If the magnification of the representation is m, the arc ds on the sphere is represented by an arc $m\,ds$ on the plane, and, if this arc makes an angle $A + \gamma$ with the x-axis, then

$$dx = m\,ds \cos (A + \gamma), \qquad dy = m\,ds \sin (A + \gamma). \qquad (5.3)$$

From the two sets of relations (5.2) and (5.3), we have

$$dx = m\,ds\,(\cos A \cos \gamma - \sin A \sin \gamma) = m \cos \phi\,(\cos \gamma \, d\psi - \sin \gamma \, d\lambda),$$
$$dy = m\,ds\,(\sin A \cos \gamma + \cos A \sin \gamma) = m \cos \phi\,(\sin \gamma \, d\psi + \cos \gamma \, d\lambda). \qquad (5.4)$$

Therefore, from (5.1) and (5.4), we have

$$\left.\begin{array}{c} \dfrac{\partial x}{\partial \psi} = \dfrac{\partial y}{\partial \lambda} = m \cos \phi \cos \gamma, \\[2ex] -\dfrac{\partial x}{\partial \lambda} = \dfrac{\partial y}{\partial \psi} = m \cos \phi \sin \gamma. \end{array}\right\} \qquad (5.5)$$

The equations (5.5) are independent of the azimuth A, so that they are true also for any other element of length at the same point subject to the same magnification m and the same rotation γ. They are therefore the conditions for any infinitesimal figure to be represented by a similar infinitesimal figure; that is, they are the conditions for conformal representation.

6. Inverse condition equations. For the inverse case of the conformal representation of the plane upon a sphere, there must again be a functional relation between two sets of isometric coordinates, and we can suppose that such a relation exists between (ψ, λ) and (x, y), whence

$$d\psi = \frac{\partial \psi}{\partial x} dx + \frac{\partial \psi}{\partial y} dy, \qquad d\lambda = \frac{\partial \lambda}{\partial x} dx + \frac{\partial \lambda}{\partial y} dy. \qquad (6.1)$$

If we consider an element of length dS at a bearing α on the plane, the components in the x and y directions respectively are

$$dx = dS \cos \alpha, \qquad dy = dS \sin \alpha. \qquad (6.2)$$

If this element is represented on the sphere by the element dS/m at an azimuth $\alpha - \gamma$, the components in meridian and parallel respectively are

$$\cos \phi \, d\psi = \frac{dS}{m} \cos (\alpha - \gamma), \qquad \cos \phi \, d\lambda = \frac{dS}{m} \sin (\alpha - \gamma). \qquad (6.3)$$

From the two sets of relations (6.2) and (6.3), we have

$$m \cos \phi \, d\psi = dS (\cos \alpha \cos \gamma + \sin \alpha \sin \gamma) = \cos \gamma \, dx + \sin \gamma \, dy,$$
$$m \cos \phi \, d\lambda = dS (\sin \alpha \cos \gamma - \cos \alpha \sin \gamma) = \cos \gamma \, dy - \sin \gamma \, dx. \qquad (6.4)$$

Therefore, from (6.1) and (6.4), we have

$$\left. \begin{aligned} \frac{\partial \psi}{\partial x} &= \frac{\partial \lambda}{\partial y} = \frac{\cos \gamma}{m \cos \phi}, \\ -\frac{\partial \lambda}{\partial x} &= \frac{\partial \psi}{\partial y} = \frac{\sin \gamma}{m \cos \phi}. \end{aligned} \right\} \qquad (6.5)$$

As before, these equations are independent of the bearing α, and are therefore the conditions for conformal representation.

7. Function establishing conformal representation. The conditions for conformal representation are expressed by the equations (5.5) and (6.5), and it remains to find a function which satisfies these equations. The expressions involving partial derivatives will be recognized as the Cauchy-Riemann equations which also express the conditions for differentiability of a function of a complex variable. That is, if the function

$$x + iy = f(u + iv) \qquad (7.1)$$

is differentiable, then by partial differentiation,

$$\left. \begin{aligned} \frac{\partial x}{\partial u} + i \frac{\partial y}{\partial u} &= \frac{d f(u + iv)}{du} = \frac{d f(u + iv)}{d(u + iv)}, \\ \frac{\partial x}{\partial v} + i \frac{\partial y}{\partial v} &= \frac{d f(u + iv)}{dv} = i \frac{d f(u + iv)}{d(u + iv)}. \end{aligned} \right\} \qquad (7.2)$$

Therefore

$$\frac{\partial x}{\partial u} + i \frac{\partial y}{\partial u} = -i \frac{\partial x}{\partial v} + \frac{\partial y}{\partial v} = \frac{d f(u + iv)}{d(u + iv)}, \qquad (7.3)$$

whence, by equating the real and the imaginary parts,

$$\left. \begin{aligned} \frac{\partial x}{\partial u} &= \frac{\partial y}{\partial v} = \text{re} \, \frac{d f(u + iv)}{d(u + iv)}, \\ -\frac{\partial x}{\partial v} &= \frac{\partial y}{\partial u} = -i \, \text{im} \, \frac{d f(u + iv)}{d(u + iv)}. \end{aligned} \right\} \qquad (7.4)$$

If we let z denote the complex variable $x + iy$ and ζ denote the complex variable $\psi + i\lambda$, then a conformal projection of the sphere is given by

$$z = f(\zeta), \qquad (7.5)$$

where $f(\zeta)$ is any analytic (differentiable) function of ζ, and hence

$$\left. \begin{aligned} \frac{\partial x}{\partial \psi} &= \frac{\partial y}{\partial \lambda} = m \cos \phi \cos \gamma = \text{re} \, \frac{d f(\zeta)}{d\zeta}, \\ -\frac{\partial x}{\partial \lambda} &= \frac{\partial y}{\partial \psi} = m \cos \phi \sin \gamma = -i \, \text{im} \, \frac{d f(\zeta)}{d\zeta}. \end{aligned} \right\} \qquad (7.6)$$

The inverse transformation is given by

$$\zeta = F(z),\tag{7.7}$$

whence

$$\left.\begin{aligned}\frac{\partial\psi}{\partial x} &= \frac{\partial\lambda}{\partial y} = \frac{\cos\gamma}{m\cos\phi} = \mathrm{re}\,\frac{\mathrm{d}F(z)}{\mathrm{d}z},\\[2mm]-\frac{\partial\lambda}{\partial x} &= \frac{\partial\psi}{\partial y} = \frac{\sin\gamma}{m\cos\phi} = -\,i\,\mathrm{im}\,\frac{\mathrm{d}F(z)}{\mathrm{d}z}.\end{aligned}\right\}\tag{7.8}$$

8. Scale coefficient and convergence. The magnification m is usually called the scale coefficient, or simply the scale, of the projection. The rotation γ is called the projection convergence, or simply the convergence; it is the angle between a projected meridian and the x-axis, or between a projected parallel and the y-axis.

From the direct equations (7.6), we have

$$m = \frac{1}{\cos\phi}\left[\left(\frac{\partial x}{\partial\psi}\right)^2 + \left(\frac{\partial y}{\partial\psi}\right)^2\right]^{\frac12}.\tag{8.1}$$

$$\tan\gamma = \frac{\partial y}{\partial\psi}\Big/\frac{\partial x}{\partial\psi},\tag{8.2}$$

and from the inverse equations (7.8), we have

$$m = \frac{1}{\cos\phi}\left[\left(\frac{\partial\psi}{\partial y}\right)^2 + \left(\frac{\partial\lambda}{\partial y}\right)^2\right]^{-\frac12},\tag{8.3}$$

$$\tan\gamma = \frac{\partial\psi}{\partial y}\Big/\frac{\partial\lambda}{\partial y}.\tag{8.4}$$

The modulus of the derivative is denoted by $|\mathrm{d}z/\mathrm{d}\zeta|$, that is

$$\left|\frac{\mathrm{d}z}{\mathrm{d}\zeta}\right| = \left[\left(\frac{\partial x}{\partial\psi}\right)^2 + \left(\frac{\partial y}{\partial\psi}\right)^2\right]^{\frac12},\tag{8.5}$$

and the conformality fails at any point where the product of $\sec\phi$ by the modulus is either zero or infinite; such a point is called a singular point of the representation.

9. Modifications for the spheroid. The radius of curvature ρ of the spheroid in the direction of the meridian at a point of latitude ϕ, and the radius of curvature ν in the direction of the decuman, are given by

$$\rho = \frac{a(1-k^2)}{(1-k^2\sin^2\phi)^{3/2}},\qquad \nu = \frac{a}{(1-k^2\sin^2\phi)^{1/2}},\tag{9.1}$$

where a is the length of the major semiaxis and k is the eccentricity of the meridian. An element of length $\mathrm{d}s$ can be resolved into components $\rho\,\mathrm{d}\phi$ along the meridian and $\nu\cos\phi\,\mathrm{d}\lambda$ along the parallel, so that

$$\mathrm{d}s^2 = \rho^2\mathrm{d}\phi^2 + \nu^2\cos^2\phi\,\mathrm{d}\lambda^2.\tag{9.2}$$

This can be put in the form (4.3) if we define the isometric latitude by

$$d\psi = \rho\, d\phi / \nu \cos\phi.$$

(9.3)

Hence by integration we get

$$\psi = \int_0^\phi \frac{(1-k^2)\,d\phi}{\cos\phi\,(1-k^2\sin^2\phi)} = \int_0^\phi \left[\frac{1}{\cos\phi} - \frac{k^2\cos\phi}{1-k^2\sin^2\phi}\right] d\phi$$

$$= \tanh^{-1}(\sin\phi) - k\tanh^{-1}(k\sin\phi),$$

(9.4)

the constant of integration being made zero by stipulating that ψ and ϕ vanish together.

The conformal representation of the spheroid can be investigated along lines precisely similar to those followed for the representation of the sphere, but using the radii of curvature ρ and ν instead of the unit radius of the sphere. In this way we arrive at the direct condition equations,

$$\left.\begin{array}{l} \dfrac{\partial x}{\partial\psi} = \dfrac{\partial y}{\partial\lambda} = m\,\nu\cos\phi\cos\gamma = \mathrm{re}\,\dfrac{df(\zeta)}{d\zeta}, \\[2ex] -\dfrac{\partial x}{\partial\lambda} = \dfrac{\partial y}{\partial\psi} = m\,\nu\cos\phi\sin\gamma = -i\,\mathrm{im}\,\dfrac{df(\zeta)}{d\zeta}, \end{array}\right\}$$

(9.5)

and the inverse condition equations,

$$\left.\begin{array}{l} \dfrac{\partial\psi}{\partial x} = \dfrac{\partial\lambda}{\partial y} = \dfrac{\cos\gamma}{m\,\nu\cos\phi} = \mathrm{re}\,\dfrac{dF(z)}{dz}, \\[2ex] -\dfrac{\partial\lambda}{\partial x} = \dfrac{\partial\psi}{\partial y} = \dfrac{\sin\gamma}{m\,\nu\cos\phi} = -i\,\mathrm{im}\,\dfrac{dF(z)}{dz}. \end{array}\right\}$$

(9.6)

The scale coefficient is given by

$$m = \frac{1}{\nu\cos\phi}\left[\left(\frac{\partial x}{\partial\psi}\right)^2 + \left(\frac{\partial y}{\partial\psi}\right)^2\right]^{\frac12} = \frac{1}{\nu\cos\phi}\left[\left(\frac{\partial\psi}{\partial y}\right)^2 + \left(\frac{\partial\lambda}{\partial y}\right)^2\right]^{-\frac12}.$$

(9.7)

10. Mercator projection. In (7.5) the simplest function is the equality

$$x + iy = \psi + i\lambda.$$

(10.1)

This gives the Mercator projection, in which

$$x = \psi = \mathrm{lam}\,\phi, \qquad y = \lambda.$$

(10.2)

Any conformal projection, defined by some function of $\psi + i\lambda$, can therefore be regarded as a conformal transformation of the Mercator projection. The transverse and oblique aspects of the same projection can therefore be defined by the same function of the transverse or the oblique Mercator coordinates.

Since

$$\partial x/\partial\psi = \partial y/\partial\lambda = 1, \qquad \partial x/\partial\lambda = \partial y/\partial\psi = 0,$$

(10.3)

the scale coefficient is given by

$$m = \sec\phi = \cosh x,$$

(10.4)

and the convergence is zero.

The projection is named from its use by Mercator 1569, although the first description was by Wright 1599 and the mathematical theory by Gregory 1668. More recent research has discovered a use of the projection by Etzlaub 1511.

The graticule of the Mercator projection is illustrated by the left-hand diagram of Fig. 1. The equator is represented by the y-axis, and the meridians are represented by straight lines orthogonal to the equator, equidistant for equal intervals of longitude. The parallels are represented by straight lines parallel to the equator, equidistant for equal intervals of isometric latitude, but with increasing separation polewards for equal intervals of geodetic latitude. The poles are represented at infinity.

11. Oblique Mercator projection. The geographic poles are two antipodal points on the sphere to which the parallels and meridians are related. We can take any other pair of antipodal points as the poles of another system of "parallels" and "meridians", thus giving us an oblique set of isometric coordinates. Such a set is defined by

$$\tanh \tfrac{1}{2}(x + iy) = i \exp(-i\phi_0) \tanh \tfrac{1}{2}(\psi + i\lambda), \qquad (11.1)$$

where (x, y) are the oblique isometric coordinates or the coordinates of an oblique Mercator projection, and ϕ_0 is the latitude of the point which is the pole of the oblique system, and is also the azimuth at which the oblique "equator" crosses the true equator. The origin of both sets of coordinates is the intersection of the two equators. If $\phi_0 = \tfrac{1}{2}\pi$, equation (11.1) reduces at once to (10.1), which defines the direct or normal Mercator projection.

The graticule of the oblique Mercator projection is illustrated by the central diagram of Fig. 1. Both meridians and parallels are symmetrical periodic curves with a quarter-period of $\tfrac{1}{2}\pi$ in the y-direction. The x-axis is an asymptote to the curve which represents the parallel of latitude ϕ_0; parallels for $\phi > \phi_0$ are closed curves around the projected position of the geographic pole; parallels for $\phi < \phi_0$ are continuous curves and have a point of inflexion where they cross the x-axis. The meridians are continuous curves and have a point of inflexion where they cross the y-axis.

Direct formulae; coordinates. Equation (11.1) can be expanded as

$$\frac{\sinh x + i \sin y}{\cosh x + \cos y} = \frac{(\sin \phi_0 + i \cos \phi_0)(\sinh \psi + i \sin \lambda)}{\cosh \psi + \cos \lambda}, \qquad (11.2)$$

from which the real and the imaginary parts can be separated as

$$\left.\begin{aligned}
\frac{\sinh x}{\cosh x + \cos y} - \frac{\sin \phi_0 \sinh \psi - \cos \phi_0 \sin \lambda}{\cosh \psi + \cos \lambda} &= g, \\
\frac{\sin y}{\cosh x + \cos y} = \frac{\cos \phi_0 \sinh \psi + \sin \phi_0 \sin \lambda}{\cosh \psi + \cos \lambda} &= h,
\end{aligned}\right\} \qquad (11.3)$$

where g and h are auxiliary symbols introduced to simplify the algebra. We now have

$$g^2 + h^2 = \frac{\cosh x - \cos y}{\cosh x + \cos y} = \frac{\cosh \psi - \cos \lambda}{\cosh \psi + \cos \lambda}, \qquad (11.4)$$

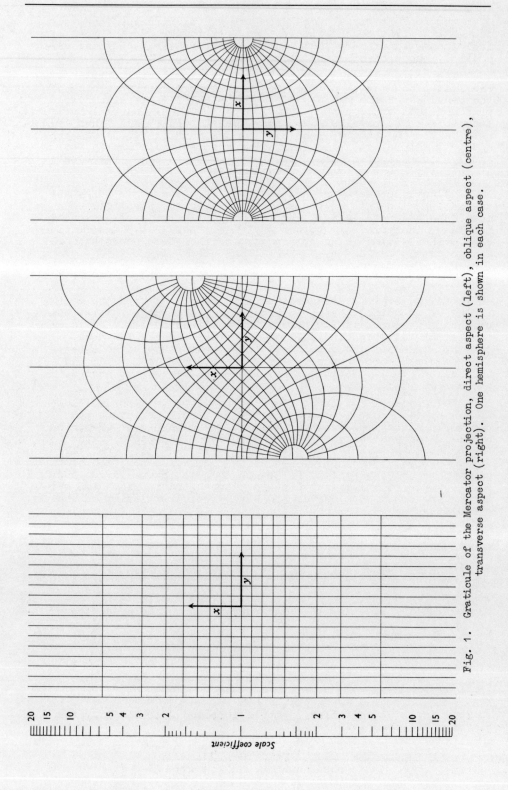

Fig. 1. Graticule of the Mercator projection, direct aspect (left), oblique aspect (centre), transverse aspect (right). One hemisphere is shown in each case.

so that solution of the equations (11.3) gives

$$\left.\begin{array}{l} \tanh x = \dfrac{2g}{1+g^2+h^2} = \dfrac{\sin\phi_0\,\sinh\psi - \cos\phi_0\,\sin\lambda}{\cosh\psi}, \\[3mm] \tan y = \dfrac{2h}{1-g^2-h^2} = \dfrac{\cos\phi_0\,\sinh\psi + \sin\phi_0\,\sin\lambda}{\cos\lambda}. \end{array}\right\} \quad (11.5)$$

Replacing hyperbolic functions of ψ by the corresponding circular functions of ϕ from (4.5), we get the coordinates of the oblique Mercator projection in the form,

$$\left.\begin{array}{l} \tanh x = \dfrac{\sin\phi_0\,\tan\phi - \cos\phi_0\,\sin\lambda}{\sec\phi}, \\[3mm] \tan y = \dfrac{\cos\phi_0\,\tan\phi + \sin\phi_0\,\sin\lambda}{\cos\lambda}. \end{array}\right\} \quad (11.6)$$

The first formula of (11.6) is suited to a mechanical calculator. For an electronic calculator, with function keys for sin, cos, tan, but not their reciprocals, it is better put in the form

$$\tanh x = \sin\phi_0\,\sin\phi - \cos\phi_0\,\cos\phi\,\sin\lambda. \quad (11.7)$$

For an electronic calculator without keys for hyperbolic functions, we can use

$$x = \tfrac{1}{2}\ln\left[(1+\tanh x)/(1-\tanh x)\right]. \quad (11.8)$$

From (11.4), the four quantities are related by the equation,

$$\cosh\psi\,\cos y = \cosh x\,\cos\lambda, \quad (11.9)$$

which can be used as a check on either the direct or the inverse computation.

Scale and convergence. Differentiation of (11.5) gives

$$\left.\begin{array}{l} \dfrac{\partial x}{\partial\psi} = \dfrac{\partial y}{\partial\lambda} = \dfrac{\sin\phi_0 + \cos\phi_0\,\sinh\psi\,\sin\lambda}{\cosh^2\psi - (\sin\phi_0\,\sinh\psi - \cos\phi_0\,\sin\lambda)^2}, \\[3mm] -\dfrac{\partial x}{\partial\lambda} = \dfrac{\partial y}{\partial\psi} = \dfrac{\cos\phi_0\,\cosh\psi\,\cos\lambda}{\cos^2\lambda + (\cos\phi_0\,\sinh\psi + \sin\phi_0\,\sin\lambda)^2}, \end{array}\right\} \quad (11.10)$$

in which the denominators of the two right-hand expressions are identical. The scale coefficient is now given by

$$m = \dfrac{\cosh\psi}{\left[\cosh^2\psi - (\sin\phi_0\,\sinh\psi - \cos\phi_0\,\sin\lambda)^2\right]^{1/2}} = \cosh x, \quad (11.11)$$

and the convergence is given by

$$\tan\gamma = \dfrac{\cosh\psi\,\cos\lambda}{\tan\phi_0 + \sinh\psi\,\sin\lambda}. \quad (11.12)$$

Change of origin. If we wish the origin to be on the straight-line meridian, we change λ and y by $\tfrac{1}{2}\pi$ so that (11.1) becomes

$$\tanh\tfrac{1}{2}[x + i(y - \tfrac{1}{2}\pi)] = i\exp(-i\phi_0)\tanh\tfrac{1}{2}[\psi + i(\lambda - \tfrac{1}{2}\pi)]. \quad (11.13)$$

By separating the real and the imaginary parts in a manner similar to that given above, or by making appropriate changes in (11.6), we now get

$$\tanh x = \frac{\sin \phi_0 \tan \phi + \cos \phi_0 \cos \lambda}{\sec \phi},$$
$$-\tan y = \frac{\sin \lambda}{\cos \phi_0 \tan \phi - \sin \phi_0 \cos \lambda}. \qquad\Bigg\} \quad (11.14)$$

In this form, the projection is sometimes called the decumanal Mercator, but it is merely an oblique Mercator with a change of origin.

The formula for $\tanh x$ can also be written

$$\tanh x = \sin \phi_0 \sin \phi + \cos \phi_0 \cos \phi \cos \lambda = \cos \sigma, \qquad (11.15)$$

where σ is the arc distance from the origin to the point (ϕ, λ) and y is the azimuth A of that arc. Thus the coordinates are

$$x = \operatorname{lam}\left(\tfrac{1}{2}\pi - \sigma\right), \qquad y = A. \qquad (11.16)$$

Inverse formulae. Equation (11.1) can also be expressed as

$$\tanh \tfrac{1}{2}(\psi + i\lambda) = -i \exp\left(i\phi_0\right) \tanh \tfrac{1}{2}(x + iy). \qquad (11.17)$$

Separation of the real and the imaginary parts gives

$$\frac{\sinh \psi}{\cosh \psi + \cos \lambda} = \frac{\cos \phi_0 \sin y + \sin \phi_0 \sinh x}{\cosh x + \cos y} = g,$$
$$\frac{\sin \lambda}{\cosh \psi + \cos \lambda} = \frac{\sin \phi_0 \sin y - \cos \phi_0 \sinh x}{\cosh x + \cos y} = h, \qquad\Bigg\} \quad (11.18)$$

from which we derive

$$g^2 + h^2 = \frac{\cosh \psi - \cos \lambda}{\cosh \psi + \cos \lambda} = \frac{\cosh x - \cos y}{\cosh x + \cos y}, \qquad (11.19)$$

and therefore

$$\sin \phi = \tanh \psi = \frac{2g}{1 + g^2 + h^2} = \frac{\cos \phi_0 \sin y + \sin \phi_0 \sinh x}{\cosh x},$$
$$\tan \lambda = \frac{2h}{1 - g^2 - h^2} = \frac{\sin \phi_0 \sin y - \cos \phi_0 \sinh x}{\cos y}. \qquad\Bigg\} \quad (11.20)$$

By rearrangement of (11.20) we get

$$\cos \phi_0 \sin y = \sin \phi \cosh x - \sin \phi_0 \sinh x,$$
$$\cos \phi_0 \sinh x = \sin \phi_0 \sin y - \tan \lambda \cos y, \qquad\Bigg\} \quad (11.21)$$

which are the equations of a parallel and of a meridian respectively.

12. Transverse Mercator projection. If $\phi_0 = 0$, equation
(11.1) becomes

$$\tan \tfrac{1}{2}i(x + iy) = -\tanh \tfrac{1}{2}(\psi + i\lambda), \qquad (12.1)$$

which defines the transverse Mercator projection with the central
meridian as the y-axis. The graticule is illustrated in the
right-hand diagram of Fig. 1, and it can be seen that the coordinate axes must be rotated positively (clockwise) through a right
angle to give the customary orientation with the central meridian
as the x-axis. This requires that $x + iy$ be multiplied by i, and
the definition of the transverse Mercator projection then becomes

$$\tan \tfrac{1}{2}(x + iy) = \tanh \tfrac{1}{2}(\psi + i\lambda). \qquad (12.2)$$

This is equivalent to

$$x + iy = \text{lam}^{-1}(\psi + i\lambda), \qquad (12.3)$$

which is a more compact definition of the projection. It also
shows at once that when $\lambda = 0$, then $y = 0$ and $x = \text{lam}^{-1}(\text{lam } \phi)$
$= \phi$; and that when $\psi = 0$, then $x = 0$ and $iy = \text{lam}^{-1}i\lambda = i \text{ lam } \lambda$,
or $y = \text{lam } \lambda$.

Direct formulae; coordinates. By separation of the real and
the imaginary parts in (12.2) we get

$$\left.\begin{array}{l} \dfrac{\sin x}{\cosh y + \cos x} = \dfrac{\sinh \psi}{\cosh \psi + \cos \lambda} = g, \\[3mm] \dfrac{\sinh y}{\cosh y + \cos x} = \dfrac{\sin \lambda}{\cosh \psi + \cos \lambda} = h, \end{array}\right\} \quad (12.4)$$

where again g and h are auxiliary symbols. We can now derive

$$g^2 + h^2 = \frac{\cosh y - \cos x}{\cosh y + \cos x} = \frac{\cosh \psi - \cos \lambda}{\cosh \psi + \cos \lambda}, \qquad (12.5)$$

whence

$$\left.\begin{array}{l} \tan x = \dfrac{2g}{1 - g^2 - h^2} = \dfrac{\sinh \psi}{\cos \lambda}, \\[3mm] \tanh y = \dfrac{2h}{1 + g^2 + h^2} = \dfrac{\sin \lambda}{\cosh \psi}. \end{array}\right\} \quad (12.6)$$

Replacing hyperbolic functions of ψ by the corresponding circular
functions of ϕ from (4.5), we finally derive the transverse Mercator coordinates in the form,

$$\tan x = \tan \phi \sec \lambda, \qquad \tanh y = \cos \phi \sin \lambda. \qquad (12.7)$$

The four quantities are related by the equations,

$$\sinh \psi \sinh y = \sin x \sin \lambda, \qquad \cosh \psi \cos x = \cosh y \cos \lambda, \qquad (12.8)$$

which can be used as checks on either the direct or the inverse
computation.

Scale and convergence. Differentiation of (12.6) gives

$$\left. \begin{array}{l} \dfrac{\partial x}{\partial \psi} = \dfrac{\partial y}{\partial \lambda} = \dfrac{\cosh \psi \cos \lambda}{\sinh^2 \psi + \cos^2 \lambda}, \\[4mm] -\dfrac{\partial y}{\partial \psi} = \dfrac{\partial x}{\partial \lambda} = \dfrac{\sinh \psi \sin \lambda}{\cosh^2 \psi - \sin^2 \lambda}, \end{array} \right\} \qquad (12.9)$$

where the two denominators are identical. The scale is therefore given by

$$m = (1 - \cos^2 \phi \sin^2 \lambda)^{-\frac{1}{2}} = \cosh y, \qquad (12.10)$$

and the convergence is given by

$$\tan \gamma = \sin \phi \tan \lambda. \qquad (12.11)$$

Inverse formulae. From equations (12.4) we can also derive

$$\left. \begin{array}{l} \sin \phi = \tanh \psi = \dfrac{2g}{1 + g^2 + h^2} = \dfrac{\sin x}{\cosh y}, \\[4mm] \tan \lambda = \dfrac{2h}{1 - g^2 - h^2} = \dfrac{\sinh y}{\cos x}. \end{array} \right\} \qquad (12.12)$$

Equations (12.12) are also the equations of a parallel and of a meridian respectively.

From (12.11) and (12.12), the convergence is given by

$$\tan \gamma = \tanh x \tan y. \qquad (12.13)$$

Change of origin. In the above investigation, the origin of coordinates is the intersection of the equator and the central meridian. If we wish the origin to be at the pole, we change y and λ by $\frac{1}{2}\pi$, and equation (12.1) becomes

$$\tan \tfrac{1}{2}\left[(\tfrac{1}{2}\pi - y) + ix \right] = -\tanh \tfrac{1}{2}\left[\psi - i(\tfrac{1}{2}\pi - \lambda) \right]. \qquad (12.14)$$

By separating the real and the imaginary parts, or by making appropriate changes in (12.7), we now get

$$\tanh x = \cos \phi \cos \lambda, \qquad \tan y = -\cot \phi \sin \lambda. \qquad (12.15)$$

The significance of the negative sign is that, for an origin at the north pole, with the initial meridian as the positive x-axis, east longitude is measured counterclockwise from that meridian.

III
Conformal Projections of the Sphere or Parts of the Sphere within a Circle

13. Conformal representation of a hemisphere within a circle. The stereographic projection represents the entire sphere conformally upon the infinite plane, but the central hemisphere is represented within a circle, and in many practical applications the projection is restricted to this hemisphere. It has the property that every circle on the sphere is represented by a circle on the projection, straight lines being admitted as circles of infinite radius. Geometrically, it may be conceived as a perspective projection onto a tangent plane from the point antipodal to the point of tangency. This gives a projection with $m = 1$ at the origin. If the projection is made upon a plane through the centre of the sphere parallel to the tangent plane, the scale is halved, so that $m = \frac{1}{2}$ at the origin. In this latter case the hemisphere is mapped within the unit circle, and this is the parent projection for many of those described later.

The stereographic is one of the oldest known projections, and is said to have been used by Hipparchus of Rhodes c. 150 B.C., but may have been known to Apollonius of Perga c. 200 B.C. The oldest extant description was by Ptolemy c. 150 A.D., and the oldest known demonstration of its conformal property was by Leadbetter 1728.

Three aspects of the stereographic projection of a hemisphere are illustrated in Fig. 2. The general or oblique aspect may have an origin at any point on the sphere; in the direct or normal aspect the origin is at a pole; and in the transverse aspect the origin is a point on the equator. As many of the projections described later are derived as transformations of the stereographic, we shall denote its coordinates by r ($= p + iq$) instead of by s ($= x + iy$). Isometric coordinates on the sphere will be denoted by ζ ($= \psi + i\lambda$).

14. Oblique aspect of stereographic projection. We first derive formulae for the oblique aspect, the most general case, and later obtain formulae for the transverse and normal aspects as limiting cases of the oblique aspect.

Direct formulae. If the origin is at a point of isometric latitude ψ_0, the projection is defined by

$$r = \cosh \psi_0 \tanh \tfrac{1}{2}(\psi_0 + \zeta) - \sinh \psi_0, \qquad (14.1)$$

from which the coordinates can be separated as

$$\left. \begin{aligned} p &= \frac{\sinh \psi - \sinh \psi_0 \cos \lambda}{\cosh \psi_0 \cosh \psi + \sinh \psi_0 \sinh \psi + \cos \lambda}, \\ q &= \frac{\cosh \psi_0 \sin \lambda}{\cosh \psi_0 \cosh \psi + \sinh \psi_0 \sinh \psi + \cos \lambda}. \end{aligned} \right\} \quad (14.2)$$

Expressed in terms of geodetic latitude by (4.5) these become

$$\left. \begin{aligned} p &= \frac{\tan \phi - \tan \phi_0 \cos \lambda}{\sec \phi_0 \sec \phi + \tan \phi_0 \tan \phi + \cos \lambda}, \\ q &= \frac{\sec \phi_0 \sin \lambda}{\sec \phi_0 \sec \phi + \tan \phi_0 \tan \phi + \cos \lambda}. \end{aligned} \right\} \quad (14.3)$$

For computation by mechanical calculator it can be noted that $\sec\phi_0 \sec\phi + \tan\phi_0 \tan\phi$ is constant along a parallel and can be tabulated initially for the required range of latitude. For an electronic calculator, the formulae are better put in the form,

$$x = \frac{\cos\phi_0 \sin\phi - \sin\phi_0 \cos\phi \cos\lambda}{1 + \sin\phi_0 \sin\phi + \cos\phi_0 \cos\phi \cos\lambda},$$

$$y = \frac{\cos\phi \sin\lambda}{1 + \sin\phi_0 \sin\phi + \cos\phi_0 \cos\phi \cos\lambda}.$$

(14.4)

Points on the initial meridian $\lambda = 0$ are given by

$$p = \tan\tfrac{1}{2}(\phi - \phi_0), \qquad q = 0,$$

(14.5)

and points on the equator $\phi = 0$ are given by

$$p = -\frac{\sin\phi_0}{\cos\phi_0 + \sec\lambda}, \qquad q = \frac{\tan\lambda}{\cos\phi_0 + \sec\lambda}.$$

(14.6)

On the boundary of a hemisphere centred on the origin, we have, from (3.6),

$$p = \cos\phi \sqrt{(\tan^2\phi + \cos^2\lambda)}, \qquad q = \cos\phi \sin\lambda,$$

(14.7)

whence $p^2 + q^2 = 1$, so that the hemisphere is contained within the unit circle.

Scale and convergence. From (14.2), partial derivatives are

$$\left.\begin{aligned}
\frac{\partial p}{\partial\psi} = \frac{\partial q}{\partial\lambda} &= \frac{\cosh\psi_0 \left[1 + \cos\lambda (\cosh\psi_0 \cosh\psi + \sinh\psi_0 \sinh\psi)\right]}{(\cosh\psi_0 \cosh\psi + \sinh\psi_0 \sinh\psi + \cos\lambda)^2}, \\
-\frac{\partial q}{\partial\psi} = \frac{\partial p}{\partial\lambda} &= \frac{\cosh\psi_0 \sin\lambda (\cosh\psi_0 \sinh\psi + \sinh\psi_0 \cosh\psi)}{(\cosh\psi_0 \cosh\psi + \sinh\psi_0 \sinh\psi + \cos\lambda)^2},
\end{aligned}\right\}$$

(14.8)

whence the scale coefficient is given by

$$m = \frac{\cosh\psi_0 \cosh\psi}{\cosh\psi_0 \cosh\psi + \sinh\psi_0 \sinh\psi + \cos\lambda},$$

(14.9)

or by

$$m = \frac{1}{1 + \sin\phi_0 \sin\phi + \cos\phi_0 \cos\phi \cos\lambda} = \tfrac{1}{2}(1 + p^2 + q^2).$$

(14.10)

The isomegeths (curves of constant scale) are circles with centres at the origin.

The convergence is given by

$$\tan\gamma = -\frac{\sin\lambda (\cosh\psi_0 \sinh\psi + \sinh\psi_0 \cosh\psi)}{1 + \cos\lambda (\cosh\psi_0 \cosh\psi + \sinh\psi_0 \sinh\psi)}$$

$$= -\frac{\sin\lambda (\sin\phi_0 + \sin\phi)}{\cos\phi_0 \cos\phi + \cos\lambda(1 + \sin\phi_0 \sin\phi)}$$

(14.11)

Inverse formulae. The definition (14.1) in inverse form is

$$\psi_0 + \zeta = 2\tanh^{-1}[(r + \sinh\psi_0)/\cosh\psi_0],\qquad(14.12)$$

which can also be written as

$$\psi_0 + \psi + i\lambda = \tanh^{-1}\frac{2\cosh\psi_0\,(p + \sinh\psi_0)}{\cosh^2\psi_0 + (p + \sinh\psi_0)^2 + q^2}$$

$$+\,i\tan^{-1}\frac{2q\cosh\psi_0}{\cosh^2\psi_0 - (p + \sinh\psi_0)^2 - q^2},\qquad(14.13)$$

and from this we get

$$\left.\begin{aligned}
\tanh\psi &= \frac{2p + \sinh\psi_0\,(1 - p^2 - q^2)}{\cosh\psi_0\,(1 + p^2 + q^2)},\\[2mm]
\tan\lambda &= \frac{2q\cosh\psi_0}{1 - p^2 - q^2 - 2p\sinh\psi_0}.
\end{aligned}\right\}\qquad(14.14)$$

Expressed in terms of geodetic latitude from (4.5) these become

$$\left.\begin{aligned}
\sin\phi &= \frac{\sin\phi_0\,(1 - p^2 - q^2) + 2p\cos\phi_0}{1 + p^2 + q^2},\\[2mm]
\tan\lambda &= \frac{2q}{\cos\phi_0\,(1 - p^2 - q^2) - 2p\sin\phi_0}.
\end{aligned}\right\}\qquad(14.15)$$

Nature of the graticule. Equations (14.15) can be rearranged to give the equation of a parallel as

$$\left(p - \frac{\cos\phi_0}{\sin\phi_0 + \sin\phi}\right)^2 + q^2 = \left(\frac{\cos\phi}{\sin\phi_0 + \sin\phi}\right)^2,\qquad(14.16)$$

and the equation of a meridian as

$$(p + \tan\phi_0)^2 + (q + \sec\phi_0\cot\lambda)^2 = \sec^2\phi_0\,\mathrm{cosec}^2\lambda,\qquad(14.17)$$

showing that parallels and meridians are represented by circles. They are two orthogonal systems of coaxal circles. The locus of meridian centres is the projection of the parallel of latitude $-\phi_0$.

When the whole sphere is mapped, if any point on the sphere is represented by r, the antipodal point is represented by $-1/\bar{r}$, where r and $\bar{r}$ are conjugate complex quantities.

15. Transverse aspect of stereographic projection. If the origin is a point on the equator, $\psi_0 = 0$, and the definition (14.1) becomes

$$r = \tanh\tfrac{1}{2}\zeta,\qquad(15.1)$$

from which the coordinates are given by

$$\left.\begin{aligned}
p &= \frac{\sinh\psi}{\cosh\psi + \cos\lambda} = \frac{\sin\phi}{1 + \cos\phi\cos\lambda},\\[2mm]
q &= \frac{\sin\lambda}{\cosh\psi + \cos\lambda} = \frac{\cos\phi\sin\lambda}{1 + \cos\phi\cos\lambda},
\end{aligned}\right\}\qquad(15.2)$$

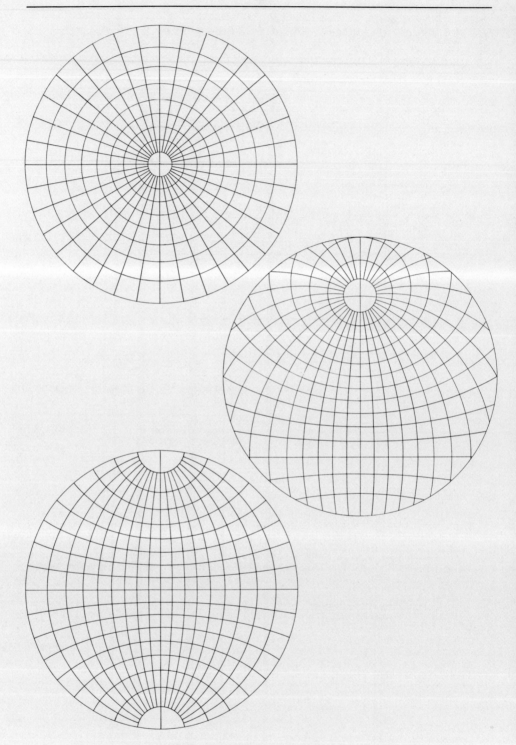

Fig. 2. Graticule of the stereographic projection of a hemisphere, direct aspect (upper), oblique aspect (centre), transverse aspect (lower).

and the scale coefficient is given by

$$m = \frac{\cosh\psi}{\cosh\psi + \cos\lambda} = \frac{1}{1 + \cos\phi\cos\lambda} = \tfrac{1}{2}(1 + p^2 + q^2). \qquad (15.3)$$

Points on the initial meridian $\lambda = 0$ are given by

$$p = \tan\tfrac{1}{2}\phi, \qquad q = 0, \qquad (15.4)$$

points on the equator $\phi = 0$ are given by

$$p = 0, \qquad q = \tan\tfrac{1}{2}\lambda, \qquad (15.5)$$

and points on the meridian $\lambda = \tfrac{1}{2}\pi$ are given by

$$p = \sin\phi, \qquad q = \cos\phi. \qquad (15.6)$$

Equation (15.1) in the inverse form,

$$\zeta = 2\tanh^{-1}r, \qquad (15.7)$$

leads at once to the inverse formulae,

$$\sin\phi = \tanh\psi = \frac{2p}{1 + p^2 + q^2}, \qquad \tan\lambda = \frac{2q}{1 - p^2 - q^2}. \qquad (15.8)$$

By rearranging equations (15.8) we obtain

$$\left.\begin{aligned} (p - \operatorname{cosec}\phi)^2 + q^2 &= \cot^2\phi, \\ p^2 + (q + \cot\lambda)^2 &= \operatorname{cosec}^2\lambda, \end{aligned}\right\} \qquad (15.9)$$

which are the equations of a parallel and of a meridian respectively and show that both curves are circles.

16. Polar aspect of stereographic projection. If the origin is a pole, $\psi_0 = \infty$, and expression (14.1) becomes indeterminate. However, on passing to exponentials, we get

$$r = \frac{\exp\zeta - \exp\psi_0}{1 + \exp\zeta\exp\psi_0} = \tan(\tan^{-1}\exp\zeta - \tan^{-1}\exp\psi_0). \qquad (16.1)$$

When $\psi_0 = \infty$, then $\tan^{-1}\exp\psi_0 = \tfrac{1}{2}\pi$, and we get

$$r = -\exp(-\zeta). \qquad (16.2)$$

The significance of the negative sign is that, with the origin at the north pole, coordinates on the initial meridian $\lambda = 0$ are measured southward.

The coordinates can be separated as

$$\left.\begin{aligned} p &= -\exp(-\psi)\cos\lambda = -\tan\tfrac{1}{2}c\cos\lambda, \\ q &= \exp(-\psi)\sin\lambda = \tan\tfrac{1}{2}c\sin\lambda, \end{aligned}\right\} \qquad (16.3)$$

the second forms following from (4.5), and the scale is given by

$$m = \tfrac{1}{2}\sec^2\tfrac{1}{2}c = \tfrac{1}{2}(1 + p^2 + q^2). \qquad (16.4)$$

Formulae for inverse computation, from (16.3), are

$$\tan \tfrac{1}{2}c = \sqrt{(p^2 + q^2)}, \qquad \tan \lambda = - q/p, \tag{16.5}$$

or
$$\tan \phi = (1 - p^2 - q^2)/2\sqrt{(p^2 + q^2)}. \tag{16.6}$$

Equations (16.5) are also the equations of a parallel and of a meridian respectively, showing that parallels are circles of radius $\tan \tfrac{1}{2}c$ with their centres at the origin, and the meridians are straight lines through the origin.

17. Lagrange conformal projection of the sphere within a circle. The Lagrange projection maps the whole sphere within a circle, with singular points at two antipodal points represented by the extremities of a diameter. If the projection is to be within the unit circle, then $m = \tfrac{1}{4}$ at the origin.

If the origin is on the equator, with a meridian as the x-axis, the definition of the Lagrange projection is

$$z = \tanh \tfrac{1}{4} \zeta, \tag{17.1}$$

which at once gives

$$x = \frac{\sinh \tfrac{1}{2}\psi}{\cosh \tfrac{1}{2}\psi + \cos \tfrac{1}{2}\lambda}, \qquad y = \frac{\sin \tfrac{1}{2}\lambda}{\cosh \tfrac{1}{2}\psi + \cos \tfrac{1}{2}\lambda}, \tag{17.2}$$

and the singular points are the geographic poles, which are represented by $x = \pm 1$, $y = 0$. This is the normal or direct aspect illustrated by the upper diagram of Fig. 3.

Points on the equator are given by

$$x = 0, \qquad y = \tan \tfrac{1}{4}\lambda, \tag{17.3}$$

points on the initial meridian $\lambda = 0$ by

$$x = \tanh \tfrac{1}{4}\psi, \qquad y = 0, \tag{17.4}$$

and points on the bounding meridian $\lambda = \pi$ by

$$x = \tanh \tfrac{1}{2}\psi, \qquad y = \operatorname{sech} \tfrac{1}{2}\psi. \tag{17.5}$$

Partial derivatives are

$$\left. \begin{aligned} \frac{\partial x}{\partial \psi} = \frac{\partial y}{\partial \lambda} &= \frac{1 + \cosh \tfrac{1}{2}\psi \, \cos \tfrac{1}{2}\lambda}{2(\cosh \tfrac{1}{2}\psi + \cos \tfrac{1}{2}\lambda)^2}, \\ -\frac{\partial y}{\partial \psi} = \frac{\partial x}{\partial \lambda} &= \frac{\sinh \tfrac{1}{2}\psi \, \sin \tfrac{1}{2}\lambda}{2(\cosh \tfrac{1}{2}\psi + \cos \tfrac{1}{2}\lambda)^2}, \end{aligned} \right\} \tag{17.6}$$

whence scale and convergence are given by

$$m = \frac{\tfrac{1}{2}\cosh \psi}{\cosh \tfrac{1}{2}\psi + \cos \tfrac{1}{2}\lambda}, \qquad \tan \gamma = - \frac{\sinh \tfrac{1}{2}\psi \, \sin \tfrac{1}{2}\lambda}{1 + \cosh \tfrac{1}{2}\psi \, \cos \tfrac{1}{2}\lambda}. \tag{17.7}$$

From (17.1) in the inverse form,

$$\zeta = 4 \tanh^{-1} z, \tag{17.8}$$

we get the formulae for inverse computation,

$$\tanh \tfrac{1}{2}\psi = \frac{2x}{1 + x^2 + y^2}, \qquad \tan \tfrac{1}{2}\lambda = \frac{2y}{1 - x^2 - y^2}, \qquad (17.9)$$

whence we also get

$$\sec \phi = \cosh \psi = \frac{(1 + x^2 + y^2)^2 + 4x^2}{(1 + x^2 + y^2)^2 - 4x^2}. \qquad (17.10)$$

Partial derivatives are

$$\left. \begin{aligned} \frac{\partial \psi}{\partial x} = \frac{\partial \lambda}{\partial y} &= \frac{4(1 - x^2 + y^2)}{(1 + x^2 + y^2)^2 - 4x^2}, \\[2mm] -\frac{\partial \psi}{\partial y} = \frac{\partial \lambda}{\partial x} &= \frac{8xy}{(1 + x^2 + y^2)^2 - 4x^2}, \end{aligned} \right\} (17.11)$$

whence, with (17.10), the scale is given by

$$m = \frac{1}{4} \cdot \frac{(1 + x^2 + y^2)^2 + 4x^2}{\sqrt{[(1 + x^2 + y^2)^2 - 4x^2]}}, \qquad (17.12)$$

and the convergence by

$$\tan \gamma = - \frac{2xy}{1 - x^2 + y^2}. \qquad (17.13)$$

Equations (17.9) can be rearranged to give

$$\left. \begin{aligned} (x - \coth \tfrac{1}{2}\psi)^2 + y^2 &= \operatorname{cosech}^2 \tfrac{1}{2}\psi, \\ x^2 + (y + \cot \tfrac{1}{2}\lambda)^2 &= \operatorname{cosec}^2 \tfrac{1}{2}\lambda, \end{aligned} \right\} (17.14)$$

which are the equations of a parallel and of a meridian respectively, and show that both are represented by circles.

The projection given by (17.1) was first described by Lambert 1772. Lagrange derived it as a special case of a more general projection with circular meridians and parallels.

18. Relation between Lagrange and stereographic projections. The Lagrange projection can also be derived by first mapping the sphere conformally upon a hemisphere by $\zeta' = \tfrac{1}{2}\zeta$, and then making a stereographic projection of that hemisphere. In the first transformation, we imagine the sphere to be cut along a meridian from pole to pole, and the surface then contracted conformally until it covers only a hemisphere, the poles remaining fixed and being singular points of the transformation. The practical result is merely to use $\tfrac{1}{2}\zeta$ instead of ζ in the formulae for the stereographic projection with centre on the equator, as can be seen by comparing (15.1) and (15.2) with (17.1) and (17.2).

Another relation can be obtained from (15.1) and (17.1), which are the formulae for the stereographic and the Lagrange projections, both with centres on the equator,

$$z_S = \tanh \tfrac{1}{2}\zeta, \qquad z_L = \tanh \tfrac{1}{4}\zeta. \qquad (18.1)$$

Hence the stereographic projection with any origin can be derived from the Lagrange projection with the same origin, or conversely, by

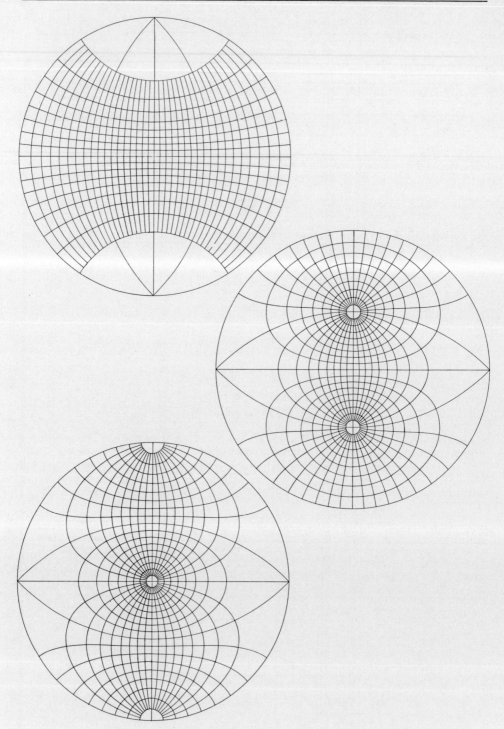

Fig. 3. Graticule of the Lagrange projection of the sphere, direct aspect (upper), transverse aspect (centre), polar aspect (lower).

$$z_S = \frac{2z_L}{1 + z_L^2}, \qquad z_L = \frac{1 - \sqrt{(1 - z_S^2)}}{z_S} \qquad (18.2)$$

19. **Oblique aspects of Lagrange projection**. The singular points of the Lagrange projection can be any pair of antipodal points on the sphere, and the boundary can be any one of the great semicircles joining the singular points. Any such oblique aspect can be derived by first computing the coordinates of an oblique Mercator projection from (11.6), changing the origin of y by a constant if necessary, and then computing the Lagrange coordinates from

$$w = \tanh \tfrac{1}{4} z. \qquad (19.1)$$

We shall consider only two cases where the coordinates can be computed simply without going through this procedure.

20. **Transverse Lagrange projection**. As the direct Lagrange projection is a conformal transformation of the Mercator, the transverse Lagrange is the same transformation of the transverse Mercator. The coordinates of the transverse Mercator have been given in (12.7) as

$$\tan x = \tan \phi / \cos \lambda, \qquad \tanh y = \sin \lambda / \sec \phi, \qquad (20.1)$$

and from these we can derive

$$\cos x = \frac{\cos \lambda}{\sqrt{(\tan^2 \phi + \cos^2 \lambda)}}, \qquad \cosh y = \frac{\sec \phi}{\sqrt{(\sec^2 \phi - \sin^2 \lambda)}}, \qquad (20.2)$$

where the two denominators are identical.

The transverse Lagrange projection is now given by

$$v + iu = \tanh \tfrac{1}{4}(y + ix) = \frac{\sinh \tfrac{1}{2}y + i \sin \tfrac{1}{2}x}{\cosh \tfrac{1}{2}y + \cos \tfrac{1}{2}x}$$

$$= \frac{\sqrt{(\cosh y - 1)} + i\sqrt{(1 - \cos x)}}{\sqrt{(\cosh y + 1)} + \sqrt{(1 + \cos x)}}. \qquad (20.3)$$

Thus, if we let

$$G = \sqrt{(\tan^2 \phi + \cos^2 \lambda)} = \sqrt{(\sec^2 \phi - \sin^2 \lambda)}, \qquad (20.4)$$

the coordinates of the transverse Lagrange projection are given by

$$u = \frac{\sqrt{(G - \cos \lambda)}}{\sqrt{(\sec \phi + G)} + \sqrt{(G + \cos \lambda)}}, \quad v = \frac{\sqrt{(\sec \phi - G)}}{\sqrt{(\sec \phi + G)} + \sqrt{(G + \cos \lambda)}}. \quad (20.5)$$

The graticule of the transverse Lagrange projection is illustrated by the central diagram of Fig. 3.

21. **Polar aspect of Lagrange projection**. If we wish the pole to be the origin of a Lagrange projection, we begin with the coordinates of a transverse Mercator projection with origin at the pole, which have been given in (12.15) as

$$\tanh x = \cos \lambda / \sec \phi, \qquad \tan y = -\sin \lambda / \tan \phi, \qquad (21.1)$$

from which we can derive

$$\cosh x = \frac{\sec\phi}{\sqrt{(\sec^2\phi - \cos^2\lambda)}}, \qquad \cos y = \frac{\tan\phi}{\sqrt{(\tan^2\phi + \sin^2\lambda)}}, \qquad (21.2)$$

where the two denominators are identical. Then, taking

$$H = \sqrt{(\sec^2\phi - \cos^2\lambda)} = \sqrt{(\tan^2\phi + \sin^2\lambda)}, \qquad (21.3)$$

by a procedure similar to that used for the transverse Lagrange, we get

$$u = \frac{\sqrt{(H - \tan\phi)}}{\sqrt{(\sec\phi + H)} + \sqrt{(H + \tan\phi)}}, \qquad v = \frac{\sqrt{(\sec\phi - H)}}{\sqrt{(\sec\phi + H)} + \sqrt{(H + \tan\phi)}}. \quad (21.4)$$

The polar Lagrange is illustrated by the lower diagram of Fig. 3.*

22. Conformal representation of certain spherical triangles on the infinite half-plane. As the infinite plane can be conformally represented within the unit circle, any part of the sphere which can be conformally represented on the infinite plane can be conformally represented within the unit circle.

In 1872 H. A. Schwarz showed that the spherical triangles into which the surface of the sphere is divided by the planes of symmetry of the regular polyhedra can be conformally represented on the infinite half-plane by algebraic transformations of the stereographic projection on the central plane so that the vertices are represented by the points, 0, 1, ∞, on the real axis. The spherical triangles occur in two sets of equal similar triangles, together completely covering the surface of the sphere, the triangles of one set being the mirror images of those in the other set. Each triangle of one set (shaded in Figs. 4 to 8) is represented on the positive half of the infinite plane, each triangle of the other set (unshaded) on the negative half. The triangles are superposed so that each one covers the entire half-plane. The real axis is slit from 1 to ∞ and from 0 to −∞, so that the regions on either side of the slits are not connected.

Schwarz's investigation is the starting point for the conformal representation of the sphere upon the five regular polyhedra. The derivations of the algebraic transformations and the preliminary theory leading to these transformations are described in Forsyth's *Theory of Functions of a Complex Variable*, Chap. XX, but the results there presented are referred to origins and axes which are inconvenient for our purpose. The results, in the form in which we require them, are therefore derived afresh in the following sections.

Coordinates on the stereographic projection are denoted by r, coordinates on the infinite plane by ζ. The method involves taking the coordinates, r_1, r_2, r_3, ..., of all points on stereographic projection corresponding to a particular vertex of the spherical triangles, and forming the product $(r_1 - r)(r_2 - r)(r_3 - r)\ldots$ If the spherical angle at the particular vertex is π/n, each such vertex occurs n times, since there are n triangles of each set with a common vertex at that point. We therefore take the product $[(r_1 - r)(r_2 - r)(r_3 - r)\ldots]^n$.

*In Wray's recognition of seven distinct aspects of a general map projection (Wray 1974), the lower diagram of Fig. 3 is first transverse, the central diagram is second transverse.

If the particular vertex is to be represented by $\xi = 0$, the product must vanish when $\xi = 0$; or, if this vertex is to be represented by $\xi = 1$, the product must vanish when $\xi = 1$; or, in the third case, the product must vanish when $\xi = \infty$. Knowing the zeros and infinities of the transforming function, we can therefore determine the function except for a constant factor which can be found from the values of r when $\xi = 1$, that is, from the relations between ξ and $1 - \xi$.

The point antipodal to the origin in the stereographic projection is represented by the point at infinity, and as $\infty - r$ has no assigned meaning, it is necessary in this case to resort to homogeneous coordinates. That is, r is replaced by r_a/r_b, where $r_b = 1$ for any point within the finite domain but $r_b = 0$ for the point at infinity, and in the latter case we can take $r_a = \pm 1$. The difference $r_n - r$, where r_n is a known value, now takes the form $r_b r_n - r_a$. For the point at infinity, we can now put $r_b = 0$, $r_a = -1$, and find that the appropriate factor in the product $(r_1 - r)(r_2 - r)(r_3 - r) \ldots$ is 1.

23. Spherical triangles for the cube. In the division of the spherical surface from which we can derive a conformal projection of the sphere upon a cube, the surface is divided by six great circles into 24 spherical triangles, each with angles $\frac{1}{2}\pi, \frac{1}{3}\pi, \frac{1}{3}\pi$. The stereographic projection of these great circles is shown in Fig. 4. (Because the triangles are isosceles, the mirror images are in this case identical with the triangles of the other set). The six points S, where two great circles intersect at angle $\frac{1}{2}\pi$, will be the centres of the faces of the cube; the four points T and the four points F, where three great circles intersect at angle $\frac{1}{3}\pi$, will be the vertices of the cube. The origin is taken at one of the points S, and the axes are the projections of the two great circles which intersect at this point, corresponding to the diagonals of the square face of the cube.

By solving the spherical triangle STF with angles $\frac{1}{2}\pi, \frac{1}{3}\pi, \frac{1}{3}\pi$, we find

$$\cos s = \frac{1}{3}, \qquad \cos t = \cos f = \frac{1}{\sqrt{3}}, \qquad (23.1)$$

and thence

$$\tan\tfrac{1}{2}s = \frac{1}{\sqrt{2}}, \qquad \tan\tfrac{1}{2}t = \tan\tfrac{1}{2}f = \frac{\sqrt{3}-1}{\sqrt{2}}. \qquad (23.2)$$

The coordinates of the six points S on stereographic projection are therefore

$$0, \quad \frac{1+i}{\sqrt{2}}, \quad -\frac{1-i}{\sqrt{2}}, \quad -\frac{1+i}{\sqrt{2}}, \quad \frac{1-i}{\sqrt{2}}, \quad \infty; \qquad (23.3)$$

the coordinates of the four points T are

$$\frac{\sqrt{3}-1}{\sqrt{2}}, \quad -\frac{\sqrt{3}-1}{\sqrt{2}}, \quad i\frac{\sqrt{3}+1}{\sqrt{2}}, \quad -i\frac{\sqrt{3}+1}{\sqrt{2}}; \qquad (23.4)$$

and the coordinates of the four points F are

$$i\frac{\sqrt{3}-1}{\sqrt{2}}, \quad -i\frac{\sqrt{3}-1}{\sqrt{2}}, \quad \frac{\sqrt{3}+1}{\sqrt{2}}, \quad -\frac{\sqrt{3}+1}{\sqrt{2}}. \qquad (23.5)$$

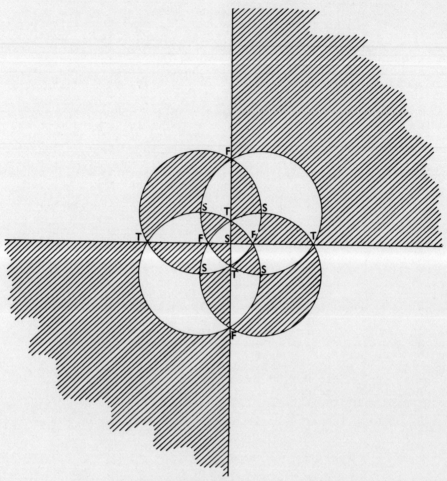

Fig. 4. Stereographic projection of spherical triangles
associated with the cube.

Four of the great circles are represented by circles of radius $\sqrt{2}$,
and the other two by straight lines through the origin; all circles
have centres at points S.

The product of the factors $(r_1 - r)(r_2 - r)(r_3 - r)\ldots$ for the
six points S is

$$- r(1 + r^4);$$
(23.6)

for the four points T the product is

$$-(1 - 2\sqrt{3}r^2 - r^4);$$
(23.7)

and for the four points F it is

$$-(1 + 2\sqrt{3}r^2 - r^4).$$
(23.8)

In the projection on the infinite plane, each point S occurs
twice, since there are two triangles of each set with a common
vertex at such a point, and each point T and each point F occurs

three times. Each vertex thus occurs 12 times, corresponding to the 12 triangles of each set.

If the vertices, S, T, F, are to be represented by 0, 1, ∞, respectively, we must have

$$\xi = 0 \text{ when } r^2(1 + r^4)^2 \text{ vanishes,}$$
$$\xi = 1 \text{ when } (1 - 2\sqrt{3}r^2 - r^4)^3 \text{ vanishes,}$$
$$\xi = \infty \text{ when } (1 + 2\sqrt{3}r^2 - r^4)^3 \text{ vanishes.}$$

We thus find

$$\xi = \frac{12\sqrt{3}r^2(1 + r^4)^2}{(1 + 2\sqrt{3}r^2 - r^4)^3}, \qquad 1 - \xi = \frac{(1 - 2\sqrt{3}r^2 - r^4)^3}{(1 + 2\sqrt{3}r^2 - r^4)^3}, \qquad (23.9)$$

which give the conformal representation on the infinite half-plane of a spherical triangle with angles $\frac{1}{2}\pi$, $\frac{1}{3}\pi$, $\frac{1}{3}\pi$, and with origin at a vertex with angle $\frac{1}{2}\pi$.

The conformal projection of the sphere upon a cube can also be derived from the same spherical triangles as are associated with the regular octahedron, which involves bisecting all the angles at points S, but the derivation from the function given above is the simpler.

24. Spherical triangles for the regular tetrahedron. The same spherical triangles as for the cube are used in deriving a conformal projection of the sphere upon a regular tetrahedron, but the stereographic projection is now referred to a different vertex as origin. The four points T are the vertices of the tetrahedron, the four points F are the centres of the faces, and the six points S are the midpoints of the edges. We take the origin at one of the points F, with the positive real axis passing through one of the adjacent points T, as shown in Fig. 5.

The coordinates of the four points F are

$$0, \quad -\sqrt{2}, \quad -\omega\sqrt{2}, \quad -\omega^2\sqrt{2}; \qquad (24.1)$$

the coordinates of the four points T are

$$\frac{1}{\sqrt{2}}, \quad \frac{\omega}{\sqrt{2}}, \quad \frac{\omega^2}{\sqrt{2}}, \quad \infty; \qquad (24.2)$$

and the coordinates of the six points S are

$$-\frac{\sqrt{3} - 1}{\sqrt{2}}, \quad \frac{\sqrt{3} + 1}{\sqrt{2}}, \qquad (24.3)$$

each multiplied by 1, ω, and ω^2. Three of the great circles are represented by circles of radius $\sqrt{3}/\sqrt{2}$, and the other three by straight lines through the origin; all circles have centres at points T.

Since $1 + \omega + \omega^2 = 0$, then for any coordinates k the product

$$(k - r)(\omega k - r)(\omega^2 k - r) = k^3 - r^3. \qquad (24.4)$$

For the four points F the product is

$$r(2\sqrt{2} + r^3); \qquad (24.5)$$

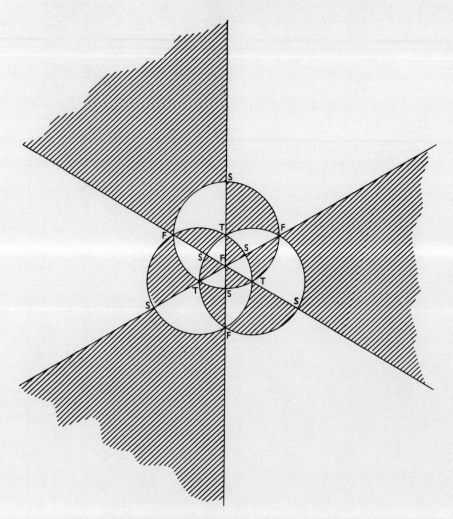

Fig. 5. Stereographic projection of spherical triangles
associated with the regular tetrahedron.

for the four points T it is

$$\frac{1 - 2\sqrt{2}r^3}{2\sqrt{2}};$$ (24.6)

and for the six points S it is

$$-(1 + 5\sqrt{2}r^3 - r^6).$$ (24.7)

The points F and the points T occur three times each, the
points S twice each. With the points F, T, S, represented by 0,
1, ∞, respectively, a conformal projection on the infinite half-
plane of a spherical triangle with angles $\frac{1}{2}\pi$, $\frac{1}{3}\pi$, $\frac{1}{3}\pi$, and with
origin at a vertex with angle $\frac{1}{3}\pi$, is given by

$$\xi = \frac{r^3(2\sqrt2 + r^3)^3}{(1 + 5\sqrt2 r^3 - r^6)^2}, \qquad 1 - \xi = \frac{(1 - 2\sqrt2 r^3)^3}{(1 + 5\sqrt2 r^3 - r^6)^2}. \qquad (24.8)$$

25. Spherical triangles for the regular octahedron. The spherical triangles associated with the regular octahedron are formed by nine great circles which divide the surface of the sphere into 48 spherical triangles, each with angles $\frac12\pi$, $\frac13\pi$, $\frac14\pi$. These are represented on stereographic projection in Fig. 6. The six points O, where four great circles intersect at angle $\frac14\pi$, are the vertices of the octahedron; the eight points C, where three great circles intersect at angle $\frac13\pi$, are the centres of the faces; and the twelve points S, where two great circles intersect at angle $\frac12\pi$, are the midpoints of the edges. The origin is taken at one of the points C, with the positive real axis passing through an adjacent point O.

By solving the spherical triangle SCO with angles $\frac12\pi$, $\frac13\pi$, $\frac14\pi$, we find

$$\cos s = \frac{1}{\sqrt3}, \qquad \cos c = \frac{1}{\sqrt2}, \qquad \cos o = \frac{\sqrt2}{\sqrt3}, \qquad (25.1)$$

and therefore

$$\tan \tfrac12 s = (\sqrt3 - 1)/\sqrt2, \qquad \tan \tfrac12 c = \sqrt2 - 1, \qquad \tan \tfrac12 o = \sqrt3 - \sqrt2. \quad (25.2)$$

The coordinates of the eight points C on stereographic projection are

$$0, \quad -\frac{1}{\sqrt2}, \quad \sqrt2, \quad \infty, \qquad (25.3)$$

each, except 0 and ∞, multiplied by 1, ω, and ω^2. The coordinates of the six points O are

$$\frac{\sqrt3 - 1}{\sqrt2}, \quad -\frac{\sqrt3 + 1}{\sqrt2}, \qquad (25.4)$$

each multiplied by 1, ω, and ω^2; and the coordinates of the twelve points S are

$$-(\sqrt3 - \sqrt2), \quad \sqrt3 + \sqrt2, \quad i, \quad -i, \qquad (25.5)$$

each multiplied by 1, ω, and ω^2. The circles on stereographic projection comprise three with radius $\sqrt3/\sqrt2$, three with radius $\sqrt3$, and the remaining three are straight lines through the origin; all circles have centres at points C.

For the points C the product $(r_1 - r)(r_2 - r)(r_3 - r)\cdots$ is

$$r(2\sqrt2 + 7r^3 - 2\sqrt2 r^6)/2\sqrt2; \qquad (25.6)$$

for the points O the product is

$$-(1 - 5\sqrt2 r^3 - r^6); \qquad (25.7)$$

and for the points S it is

$$-(1 + 22\sqrt2 r^3 + 22\sqrt2 r^9 - r^{12}). \qquad (25.8)$$

Fig. 6. Stereographic projection of spherical triangles
associated with the regular octahedron.

The points C occur three times each, the points O four times
each, and the points S twice each, corresponding to the 24 triangles of each set, each of which is to be represented on the infinite half-plane. If the points C, O, S, are to be represented
by 0, 1, ∞, respectively, the conformal representation of the
spherical triangle with angles $\frac{1}{2}\pi$, $\frac{1}{3}\pi$, $\frac{1}{4}\pi$ is given by

$$\left. \begin{aligned} \xi &= \frac{4r^3(2\sqrt{2} + 7r^3 - 2\sqrt{2}r^6)^3}{(1 + 22\sqrt{2}r^3 + 22\sqrt{2}r^9 - r^{12})^2}, \\ 1 - \xi &= \frac{(1 - 5\sqrt{2}r^3 - r^6)^4}{(1 + 22\sqrt{2}r^3 + 22\sqrt{2}r^9 - r^{12})^2}. \end{aligned} \right\} \quad (25.9)$$

26. Spherical triangles for the regular icosahedron. For the regular icosahedron, the surface of the sphere is divided by 15 great circles into 120 spherical triangles, each with angles $\frac{1}{2}\pi$, $\frac{1}{3}\pi$, $\frac{1}{5}\pi$. These are represented on stereographic projection in Fig. 7. The twelve points I, where five great circles intersect at angle $\frac{1}{5}\pi$, are the vertices of the icosahedron; the 20 points D, where three great circles intersect at angle $\frac{1}{3}\pi$, are the centres of the faces; and the 30 points S, where two great circles intersect at angle $\frac{1}{2}\pi$, are the midpoints of the edges. The origin is taken at one of the points D, with the positive real axis passing through an adjacent point I.

Solving the spherical triangle, we find that if l, m, n are the sides opposite the angles $\frac{1}{2}\pi$, $\frac{1}{3}\pi$, $\frac{1}{5}\pi$ respectively, then

$$\cos l = \frac{\surd(5+2\surd5)}{\surd3\surd5}, \qquad \cos m = \frac{\surd(5+\surd5)}{\surd2\surd5}, \qquad \cos n = \frac{\surd5+1}{2\surd3}, \qquad (26.1)$$

whence

$$\tan l = 3 - \surd5, \qquad \tan m = \tfrac{1}{2}(\surd5 - 1), \qquad \tan n = \tfrac{1}{2}(3 - \surd5). \qquad (26.2)$$

The coordinates of the 20 points D on stereographic projection are

$$0, \quad -\tfrac{1}{2}(3-\surd5), \quad \tfrac{1}{2}(3+\surd5), \quad -\tfrac{1}{4}(\surd5 - i\surd3),$$
$$-\tfrac{1}{4}(\surd5 + i\surd3), \quad \tfrac{1}{2}(\surd5 - i\surd3), \quad \tfrac{1}{2}(\surd5 + i\surd3), \quad \infty, \qquad (26.3)$$

each, except 0 and ∞, multiplied by 1, ω, and ω^2. The coordinates of the 12 points I are

$$\tan\tfrac{1}{2}l, \quad -\cot\tfrac{1}{2}l, \quad -\tan(\tfrac{1}{2}l + n), \quad \cot(\tfrac{1}{2}l + n), \qquad (26.4)$$

that is,

$$\tfrac{1}{4}[\surd2\surd3\surd(5+\surd5) - 3 - \surd5], \quad -\tfrac{1}{4}[\surd2\surd3\surd(5+\surd5) + 3 + \surd5],$$
$$-\tfrac{1}{4}[\surd2\surd3\surd(5-\surd5) - 3 + \surd5], \quad \tfrac{1}{4}[\surd2\surd3\surd(5-\surd5) + 3 - \surd5], \qquad (26.5)$$

each multiplied by 1, ω, and ω^2. The coordinates of the 30 points S are

$$-\tfrac{1}{2}(\surd3-1)(\surd5-\surd3), \quad \tfrac{1}{2}(\surd3+1)(\surd5+\surd3), \quad \tfrac{1}{2}(\surd3+1)(\surd5-\surd3),$$
$$-\tfrac{1}{2}(\surd3-1)(\surd5+\surd3), \quad \tfrac{1}{4}(\surd3-1)(\surd5+i\surd3), \quad \tfrac{1}{4}(\surd3-1)(\surd5-i\surd3),$$
$$-\tfrac{1}{4}(\surd3+1)(\surd5+i\surd3), \quad -\tfrac{1}{4}(\surd3+1)(\surd5-i\surd3), \quad i, \quad -i, \qquad (26.6)$$

each multiplied by 1, ω, and ω^2. On stereographic projection, three circles have radius $\tfrac{1}{2}\surd3(\surd5-1)$, six have radius $\surd3$, three have radius $\tfrac{1}{2}\surd3(\surd5+1)$, and the remaining three are straight lines through the origin; all circles have centres at points D.

The product of the factors for the points D is

$$\tfrac{1}{8}r(8 + 57\surd5r^3 - 228r^6 + 494\surd5r^9 + 228r^{12} + 57\surd5r^{15} - 8r^{18}); \qquad (26.7)$$

for the points I the product is

$$1 - 11\surd5r^3 - 33r^6 + 11\surd5r^9 + r^{12}; \qquad (26.8)$$

and for the points S it is

Fig. 7. Stereographic projection of spherical triangles
associated with the regular icosahedron.

$$\tfrac{1}{8}(8 + 580\sqrt{5}r^3 + 3915r^6 + 20010\sqrt{5}r^9 - 190095r^{12} - 190095r^{18}$$
$$- 20010\sqrt{5}r^{21} + 3915r^{24} - 580\sqrt{5}r^{27} + 8r^{30}). \qquad (26.9)$$

Each point D occurs three times, each point I five times, and
each point S twice, corresponding to the 60 triangles of each set.
The spherical triangle with angles $\tfrac{1}{2}\pi$, $\tfrac{1}{2}\pi$, $\tfrac{1}{5}\pi$ is therefore conformally represented upon the infinite half-plane, with the vertices
D, I, S represented by 0, 1, ∞ respectively, by

$$\xi = \frac{25\sqrt{5}r^3(8 + 57\sqrt{5}r^3 - 228r^6 + 494\sqrt{5}r^9 + 228r^{12} + 57\sqrt{5}r^{15} - 8r^{18})^3}{(8 + 580\sqrt{5}r^3 + 3915r^6 + 20010\sqrt{5}r^9 - 190095r^{12} - 190095r^{18}}$$
$$\qquad\qquad - 20010\sqrt{5}r^{21} + 3915r^{24} - 580\sqrt{5}r^{27} + 8r^{30})^2 ,$$

$$1 - \xi = \frac{64(1 - 11\sqrt{5}r^3 - 33r^6 + 11\sqrt{5}r^9 + r^{12})^5}{(8 + 580\sqrt{5}r^3 + 3915r^6 + 20010\sqrt{5}r^9 - 190095r^{12} - 190095r^{18} \\ - 20010\sqrt{5}r^{21} + 3915r^{24} - 580\sqrt{5}r^{27} + 8r^{30})^2}. \qquad (26.10)$$

27. Spherical triangles for the regular dodecahedron. For the regular dodecahedron we use the same spherical triangles as for the icosahedron, but the stereographic projection is referred to a different vertex as origin. The 20 points D are the vertices of the dodecahedron, the 12 points I are the centres of the faces, and the 30 points S are the midpoints of the edges. We take the origin at one of the points I, with the positive real axis passing through an adjacent point D, as shown in Fig. 8.

Let $\omega^5 = 1$; then

$$\omega = \cos \tfrac{2}{5}\pi + i \sin \tfrac{2}{5}\pi = -[(\sqrt{5} - 1) + i\sqrt{2}\sqrt{(5 + \sqrt{5})}]. \qquad (27.1)$$

Since $1 + \omega + \omega^2 + \omega^3 + \omega^4 = 0$, then for any coordinates k the product

$$\prod_{n=0}^{4} (\omega^n k - r) = k^5 - r^5. \qquad (27.2)$$

The coordinates of the 12 points I on stereographic projection are

$$0, \quad -\tfrac{1}{2}(\sqrt{5} - 1), \quad \tfrac{1}{2}(\sqrt{5} + 1), \quad \infty, \qquad (27.3)$$

each, except 0 and ∞, multiplied by ω^n, where $n = 0, 1, 2, 3,$ and 4. The coordinates of the 20 points D are

$$\tan \tfrac{1}{2}l, \quad -\cot \tfrac{1}{2}l, \quad \tan(\tfrac{1}{2}l + n), \quad -\cot(\tfrac{1}{2}l + n), \qquad (27.4)$$

that is,

$$\tfrac{1}{4}[\sqrt{2}\sqrt{3}\sqrt{(5 + \sqrt{5})} - 3 - \sqrt{5}], \quad -\tfrac{1}{4}[\sqrt{2}\sqrt{3}\sqrt{(5 + \sqrt{5})} + 3 + \sqrt{5}],$$
$$\tfrac{1}{4}[\sqrt{2}\sqrt{3}\sqrt{(5 - \sqrt{5})} - 3 + \sqrt{5}], \quad -\tfrac{1}{4}[\sqrt{2}\sqrt{3}\sqrt{(5 - \sqrt{5})} + 3 - \sqrt{5}], \qquad (27.5)$$

each multiplied by ω^n; and the coordinates of the 30 points S are

$$-\tan \tfrac{1}{2}m, \quad \cot \tfrac{1}{2}m, \quad \tan \tfrac{1}{2}(l + n), \quad -\cot \tfrac{1}{2}(l + n), \quad i, \quad -i, \qquad (27.6)$$

that is,

$$-\tfrac{1}{2}[\sqrt{2}\sqrt{(5 + \sqrt{5})} - \sqrt{5} - 1], \quad \tfrac{1}{2}[\sqrt{2}\sqrt{(5 + \sqrt{5})} + \sqrt{5} + 1],$$
$$\tfrac{1}{2}[\sqrt{2}\sqrt{(5 - \sqrt{5})} - \sqrt{5} + 1], \quad -\tfrac{1}{2}[\sqrt{2}\sqrt{(5 - \sqrt{5})} + \sqrt{5} - 1], \quad i, \quad -i, \qquad (27.7)$$

each multiplied by ω^n. On stereographic projection, five circles have radius $\sqrt{(5 - \sqrt{5})}/\sqrt{2}$, five have radius $\sqrt{(5 + \sqrt{5})}/\sqrt{2}$, and the remaining five are straight lines through the origin; all circles have centres at points I.

The product of the factors for the points I is

$$r(1 + 11r^5 - r^{10}); \qquad (27.8)$$

for the points D the product is

$$1 - 228r^5 + 494r^{10} + 228r^{15} + r^{20}; \qquad (27.9)$$

and for the points S it is

Fig. 8. Stereographic projection of spherical triangles
associated with the regular dodecahedron.

$$1 + 522r^5 - 10005r^{10} - 10005r^{20} - 522r^{25} + r^{30}. \qquad (27.10)$$

Each point I occurs five times, each point D three times, and
each point S twice. The spherical triangle is therefore conformally represented upon the infinite half-plane, with vertices I, D,
S represented by 0, 1, ∞ respectively, by

$$\left.\begin{array}{l} \xi = \dfrac{1728r^5(1 + 11r^5 - r^{10})^5}{(1 + 522r^5 - 10005r^{10} - 10005r^{20} - 522r^{25} + r^{30})^2}, \\[2ex] 1 - \xi = \dfrac{(1 - 228r^5 + 494r^{10} + 228r^{15} + r^{20})^3}{(1 + 522r^5 - 10005r^{10} - 10005r^{20} - 522r^{25} + r^{30})^2}. \end{array}\right\} (27.11)$$

28. Spherical triangles for a double pyramid. A conformal projection of the sphere upon a regular double pyramid of $2n$ sides can be derived from the division of the surface into equal biquadrantal spherical triangles by one great circle (e.g. the equator) and the secondaries to that great circle (the meridians) intersecting at angles (differences of longitude) of $2\pi/n$. In the stereographic projection, with origin at one of the poles, the equator is represented by the unit circle and the meridians by equally-inclined radiating straight lines.

The coordinates of the two poles are

$$0, \quad \infty. \tag{28.1}$$

The other vertices are in two groups of n points each, occurring alternately around the unit circle. The coordinates of the n vertices, of which the first is taken on the real axis, are

$$1, \quad \exp(2i\pi/n), \quad \exp(4i\pi/n), \quad \ldots, \quad \exp[(2n-2)i\pi/n)], \tag{28.2}$$

i.e., $\exp(2ki\pi/n)$ for $k = 0, 1, 2, \ldots, (n-1)$. The coordinates of the other n vertices are

$$\exp(i\pi/n), \quad \exp(3i\pi/n), \quad \exp(5i\pi/n), \quad \ldots, \quad \exp[(2n-1)i\pi/n)], \tag{28.3}$$

i.e., $\exp[(2k+1)i\pi/n]$ for $k = 0, 1, 2, \ldots, (n-1)$.

The product of the factors for the two poles is $-r$. For the first group of n vertices, it is $\pm(1-r^n)$, the sign being positive when n is odd, negative when n is even. For the second group of n vertices, the product is $\mp(1+r^n)$, the sign being negative when n is odd, positive when n is even.

The poles occur n times each, the other vertices twice each. The conformal projection of each spherical triangle upon the infinite half-plane, with origin at one of the poles, is therefore given by

$$\xi = \frac{4r^n}{(1+r^n)^2}, \qquad 1-\xi = \frac{(1-r^n)^2}{(1+r^n)^2}. \tag{28.4}$$

If n is odd, no vertex lies on the imaginary axis. If we wish to orient the projection so that one vertex does lie on the imaginary axis, we multiply all the coordinates in the above investigation by i. The only case we shall have occasion to use later is $n = 3$, one group of vertices being at i, $i\omega$, $i\omega^2$, and the other group being at $-i$, $-i\omega$, $-i\omega^2$. The conformal projection on the infinite half-plane is given by

$$\xi = -\frac{4ir^3}{(i-r^3)^2}, \qquad 1-\xi = \frac{(i+r^3)^2}{(i-r^3)^2}. \tag{28.5}$$

29. Conformal representation of the infinite plane within the unit circle. Algebraic transformations will also effect the conformal representation of the unit circle upon the infinite plane, or conversely will represent the infinite plane within the unit circle. Thus if z is the circle and w is the infinite plane, the transformations,

$$w = \frac{2z}{1+z^2}, \qquad z = \frac{1-\sqrt{(1-w^2)}}{w} \tag{29.1}$$

will represent the unit circle upon the infinite plane, or the
converse, so that the points, +1, 0, -1, in one figure are represented by the same points, +1, 0, -1, in the other figure. The
circumference of the circle is represented by the parts of the
real axis in the w-plane extending from +1 to ∞ and from -1 to $-\infty$,
and the positive and negative halves of the area within the circle
are represented respectively by the positive and negative halves
of the infinite plane. Thus the infinite plane is regarded as cut
along the real axis from +1 to $+\infty$ and from -1 to $-\infty$, so that the
areas on either side of the axis in these regions are not connected.

The relations (29.1) apply to the transformation of the
Lagrange projection into the stereographic, or conversely, as in
(18.2). In this case, the two parts of the circumference of the
bounding circle of the Lagrange projection are two positions of
the same meridian, so that the cuts along the real axis mentioned
above do not apply when the projection is that of the whole sphere.

The transformations,

$$w = \frac{4s}{(1+s)^2}, \qquad s = \frac{[1-\sqrt{(1-w)}]^2}{w}, \qquad (29.2)$$

represent the unit circle upon the infinite plane, or the converse,
so that the points, +1, 0, -1, of the circle are represented respectively by the points, +1, 0, ∞, of the infinite plane. In this
case the plane is cut only along the positive part of the real axis
from +1 to ∞, which represents the circumference of the circle.

30. Conformal representation of the unit circle within a sector of the unit circle. The transformations,

$$w = s^{1/n}, \qquad s = w^n, \qquad (30.1)$$

will conformally represent the unit circle s within a sector w of
the unit circle with angle $2\pi/n$, or will represent any sector with
angle θ within a sector with angle θ/n, so that the points, 0, +1,
in each figure will still correspond. The same transformations
will conformally contract the infinite plane into an infinite
triangle bounded by the real axis and by a straight line through
the origin making an angle $2\pi/n$ with the real axis, or will
expand the infinite triangle into the infinite plane.

31. Conformal representation of spherical triangles within
sectors of the unit circle. The above transformations can be used
in combination to represent the spherical triangles associated
with the regular polyhedra within sectors of the unit circle, such
that all the triangles which together make up one face of the polyhedron will together make up the unit circle.

In the case of the cube, a spherical triangle with angle $\frac{1}{2}\pi$
at the origin (being one-quarter of that area of the sphere which
is to be represented by one square face of the cube) has been
represented upon the infinite half-plane with vertices at 0, 1, ∞,
by the function ξ in (23.9) above. The infinite half-plane can be
represented upon the quarter-plane by taking the square root of
this function, $\sqrt{\xi}$. The quarter-plane can now be represented within
a quadrant of the unit circle, or the whole plane within the whole
circle, by the transformation,

$$s = (1-\sqrt{1-\xi})/\sqrt{\xi}. \qquad (31.1)$$

In the cases of the tetrahedron, octahedron, and icosahedron, a spherical triangle with angle $\frac{1}{3}\pi$ at the origin (in each case being one-sixth of that area of the sphere which is to be represented by one equilateral triangular face of the polyhedron) has been represented upon the infinite half-plane by the various functions ξ given in (24.8), (25.9), and (26.10) above. In each case this triangle can be represented within a sector of the unit circle with angle $\frac{1}{3}\pi$ by the transformation,

$$z = [(1-\sqrt{1-\xi})^2/\xi]^{1/3}. \qquad (31.2)$$

Similarly, for the dodecahedron, where the angle of the spherical triangle at the origin is $\frac{1}{5}\pi$ (and the triangle is one-tenth of that area of the sphere which is to be represented by one pentagonal face of the dodecahedron), we use the function ξ in (27.11) and the transformation,

$$z = [(1-\sqrt{1-\xi})^2/\xi]^{1/5}. \qquad (31.3)$$

In each case, the area of the sphere corresponding to one face of the polyhedron is now represented by the area of the unit circle.

IV
Conformal Projections Based on Dixon Elliptic Functions

32. Conformal projection of a circle within an equilateral triangle.

The *transformation* sm w = z. For the case n = 3, Schwarz's integral (3.1) can be expressed as

$$\text{sm}\, w = z,\qquad\qquad (32.1)$$

by which the interior of the unit circle z is conformally represented by the interior of the equilateral triangle w. Adams did not mention this fact, regarding his projections as being contained within a rhombus with angles of 60° and 120°, which is the period parallelogram of the Dixon functions. The long diagonal of this parallelogram is a side of the equilateral triangle. The relation of the two figures is shown in Fig. 9. The origin of coordinates is the centroid of the equilateral triangle (or a vertex of the rhombus), and there are singular points at the three vertices, K, ωK, and $\omega^2 K$.

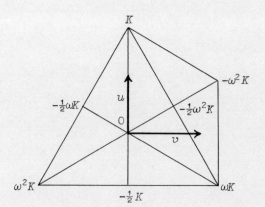

Fig. 9. Relation between equilateral triangle and rhombus in the transformation sm w = z.

For certain points on the boundary, we find

$$\left.\begin{array}{llll}
\text{sm}\, K & = 1 & \text{sm}\,(-\tfrac{1}{2}K) & = \exp i\pi \\
\text{sm}\,(-i\omega K/\sqrt3) & = \exp(i\pi/9) & \text{sm}\,(i\omega K/\sqrt3) & = \exp(11i\pi/9) \\
\text{sm}\,(-\tfrac{1}{2}\omega^2 K) & = \exp(i\pi/3) & \text{sm}\,\omega^2 K & = \exp(4i\pi/3) \\
\text{sm}\,(iK/\sqrt3) & = \exp(5i\pi/9) & \text{sm}\,(-iK/\sqrt3) & = \exp(13i\pi/9) \\
\text{sm}\,\omega K & = \exp(2i\pi/3) & \text{sm}\,(-\tfrac{1}{2}\omega K) & = \exp(5i\pi/3) \\
\text{sm}\,(-i\omega^2 K/\sqrt3) & = \exp(7i\pi/9) & \text{sm}\,(i\omega^2 K/\sqrt3) & = \exp(17i\pi/9)
\end{array}\right\}\quad(32.2)$$

and it can be noted that each sextant of the unit circle is represented by a sextant of the equilateral triangle.

Conformal projections of the sphere, or parts of the sphere, within an equilateral triangle can be derived by first mapping the area conformally within the unit circle, and then applying this transformation.

Computation of coordinates by series. For points not too distant from the origin, computation of coordinates is most easily done by series. Thus, we have

$$w = \mathrm{sm}^{-1}z = z + \frac{2}{3}\cdot\frac{1}{4}z^4 + \frac{2.5}{3.6}\cdot\frac{1}{7}z^7 + \frac{2.5.8}{3.6.9}\cdot\frac{1}{10}z^{10} + \cdots, \quad (32.3)$$

or

$$
\begin{aligned}
w = \quad & z & &+0\cdot017\ 274\,z^{22}\\
&+0\cdot166\ 667\,z^4 & &+0\cdot014\ 568\,z^{25}\\
&+0\cdot079\ 365\,z^7 & &+0\cdot012\ 525\,z^{28}\\
&+0\cdot049\ 383\,z^{10} & &+0\cdot010\ 936\,z^{31}\\
&+0\cdot034\ 821\,z^{13} & &+0\cdot009\ 669\,z^{34}\\
&+0\cdot026\ 406\,z^{16} & &+\cdots,\\
&+0\cdot021\ 001\,z^{19} & & \quad (32.4)
\end{aligned}
$$

which gives 6-figure accuracy for $|z| \not> 0\cdot75$. The inverse series is

$$z = \mathrm{sm}\,w = w - \frac{1}{6}w^4 + \frac{2}{63}w^7 - \frac{13}{2268}w^{10} + \cdots, \quad (32.5)$$

or

$$
\begin{aligned}
z = \quad & w & &-0\cdot000\ 189\,w^{16}\\
&-0\cdot166\ 667\,w^4 & &+0\cdot000\ 034\,w^{19}\\
&+0\cdot031\ 746\,w^7 & &-0\cdot000\ 006\,w^{21}\\
&-0\cdot005\ 732\,w^{10} & &+0\cdot000\ 001\,w^{25}\\
&+0\cdot001\ 040\,w^{13} & &-\cdots, \quad (32.6)
\end{aligned}
$$

which gives 6-figure accuracy for $|w| \not> 1$, which is about 72% of the area of the equilateral triangle. The inverse series can be used to check coordinates computed from the direct series, and can also be used to compute by iterations the coordinates of points beyond the range of the direct series.

Separation of the real and the imaginary parts of $\mathrm{sm}\,w$. For points beyond the range of the series, it is necessary to separate the real and the imaginary parts of $\mathrm{sm}\,w$. If a numerical value of $\mathrm{sm}\,w$ is known, we can find the value of $\mathrm{cm}\,w$ from the relation $\mathrm{cm}^3w = 1 - \mathrm{sm}^3w$. As cm^3w is a complex quantity, it can be expressed in the form $r^3(\cos 3\theta + i \sin 3\theta)$, whence one value of the cube root is $r(\cos\theta + i\sin\theta)$. It should be noted that there are three possible values of a cube root, and the value first found may need to be multiplied by ω or by ω^2; only one value of $\mathrm{cm}\,w$ may satisfy the conditions of the problem. In the computations for the projections in this study, cases where the first value of $\mathrm{cm}\,w$ was not the correct one were rare.

When $\mathrm{sm}\,w$ and $\mathrm{cm}\,w$ are known, we can find w by an algebraic or by a trigonometric method. Thus, in the first case, let

$$\left.\begin{aligned}
\mathrm{sm}\,w &= x + iy, & \mathrm{sm}\,\bar{w} &= x - iy,\\
\mathrm{cm}\,w &= k - il, & \mathrm{cm}\,\bar{w} &= k + il.
\end{aligned}\right\} \quad (32.7)$$

Then

$$\mathrm{sm}\,(w + \bar{w}) = \frac{\mathrm{sm}^2w\,\mathrm{cm}\,\bar{w} - \mathrm{cm}\,w\,\mathrm{sm}^2\bar{w}}{\mathrm{sm}\,w\,\mathrm{cm}^2\bar{w} - \mathrm{cm}^2w\,\mathrm{sm}\,\bar{w}}, \quad (32.8)$$

or
$$\text{sm } 2u \;=\; \frac{2xyk \;+\; l(x^2 - y^2)}{2klx \;+\; y(k^2 - l^2)}. \tag{32.9}$$

Also
$$\text{cm}\,(w - \bar{w}) \;=\; \frac{\text{cm}^2 w \;\text{cm}\,\bar{w} \;-\; \text{sm}\,w \;\text{sm}^2\bar{w}}{\text{cm}\,w \;\text{cm}^2\bar{w} \;-\; \text{sm}^2 w \;\text{sm}\,\bar{w}}, \tag{32.10}$$

or
$$\text{cm } 2iv \;=\; \frac{1 + \omega \,\text{sm}^3\,(2v/\sqrt{3})}{1 + \omega^2 \,\text{sm}^3\,(2v/\sqrt{3})}$$

$$\;=\; \frac{(x^2 + y^2)(x - iy) - (k^2 + l^2)(k - il)}{(x^2 + y^2)(x + iy) - (k^2 + l^2)(k + il)}, \tag{32.11}$$

whence
$$\text{sm}^3 \frac{2v}{\sqrt{3}} \;=\; \frac{2\,[\,l(k^2 + l^2) - y(x^2 + y^2)\,]}{(\sqrt{3}x - y)(x^2 + y^2) - (\sqrt{3}k - l)(k^2 + l^2)}. \tag{32.12}$$

An alternative method was given by Adams, although his results and his derivation of them are more complicated than they need be. Let

$$\begin{aligned} \text{sm}\,w &= Se^{is}, & \text{sm}\,\bar{w} &= Se^{-is}, \\ \text{cm}\,w &= Ce^{-ic}, & \text{cm}\,\bar{w} &= Ce^{ic}. \end{aligned} \left.\right\} \tag{32.13}$$

Then, from (32.8),

$$\text{sm}\,(w + \bar{w}) \;=\; \frac{S^2 C\,[\,e^{i(2s + c)} - e^{-i(2s + c)}\,]}{SC^2\,[\,e^{i(s + 2c)} - e^{-i(s + 2c)}\,]}, \tag{32.14}$$

or
$$\text{sm } 2u \;=\; \frac{S\,\sin\,(2s + c)}{C\,\sin\,(s + 2c)}. \tag{32.15}$$

Also, from (32.10),

$$\text{cm}\,(w - \bar{w}) \;=\; \text{cm } 2iv \;=\; \frac{1 + \omega \,\text{sm}^3\,(2v/\sqrt{3})}{1 + \omega^2 \,\text{sm}^3\,(2v/\sqrt{3})} \;=\; \frac{C^3 e^{-ic} - S^3 e^{-is}}{C^3 e^{ic} - S^3 e^{is}}, \tag{32.16}$$

whence

$$\text{sm}^3 \frac{2v}{\sqrt{3}} \;=\; \frac{S^3(e^{is} - e^{-is}) - C^3(e^{ic} - e^{-ic})}{S^3(\omega^2 e^{-is} - \omega e^{is}) - C^3(\omega^2 e^{-ic} - \omega e^{ic})}$$

$$\;=\; \frac{S^3 \sin s - C^3 \sin c}{C^3 \sin\,(\tfrac{1}{3}\pi - c) - S^3 \sin\,(\tfrac{1}{3}\pi - s)}. \tag{32.17}$$

Tables of Dixon functions. Table 2 gives 12-decimal values of sm u and cm u for every hundredth of the range from 0 to K, and Table 3 gives 5-decimal values for every tenth of the complete period.

Tables 4 to 10 are provided for finding u from sm $2u$ and v from sm$^3(2v/\sqrt{3})$, using formulae (32.9) and (32.12) or (32.15) and (32.17). These are designed for giving 6-decimal accuracy, but they are inadequate when sm $2u$ is close to 1 or when v is very small.

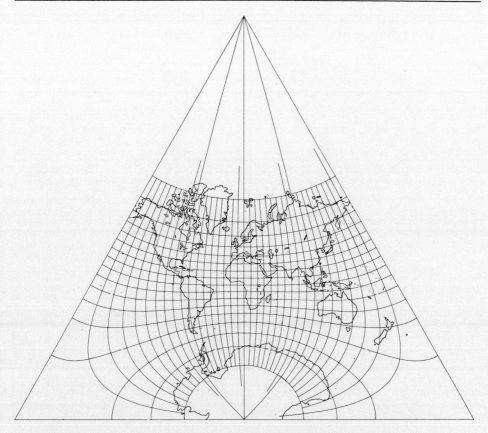

Fig. 10. Projection of the sphere within an equilateral triangle.

33. Projection of the sphere within an equilateral triangle (Cox 1935, Magis 1938). Adams mentioned the possibility of mapping the sphere conformally within an equilateral triangle, but stated that "such a projection would not be especially important for geographic purposes, so that no computations were made for it." Coordinates of the projection were published by Magis 1938, but they are very inaccurate, no doubt because he followed Adams's complicated methods of computation.

The projection can be simply expressed by

$$\operatorname{sm} w = \tanh \tfrac{1}{4}\zeta, \qquad (33.1)$$

(also noted by Gougenheim 1953), where $\tanh \tfrac{1}{4}\zeta$ is the Lagrange projection of the sphere within a circle, with origin on the equator and with $m = \tfrac{1}{4}$ at the origin. Thus, we first compute the coordinates of the Lagrange projection from (17.2), and then compute w as described in Sec. 32 above.

Scale. From (33.1) we have

$$\operatorname{cm}^2 w \, \mathrm{d}w = \tfrac{1}{4} \operatorname{sech}^2 \tfrac{1}{4}\zeta \, \mathrm{d}\zeta = \tfrac{1}{4}(1 - \operatorname{sm}^2 w) \, \mathrm{d}\zeta, \qquad (33.2)$$

whence
$$\frac{\mathrm{d}w}{\mathrm{d}\zeta} = \frac{1 - \operatorname{sm}^2 w}{4 \operatorname{cm}^2 w}. \qquad (33.3)$$

There are four singular points, at each of the three vertices, and at the midpoint of one of the sides (in this case, at $w = -\frac{1}{2}K$).

The scale of the projection is given by $\sec\phi\,|dw/d\zeta|$, again with $m = \frac{1}{4}$ at the origin.

Nature of the projection. Fig. 10 illustrates a projection in which the north pole is at one vertex, and the south pole is at the midpoint of the opposite side, both poles being singular points. The other two singular points (two representations of the same point on the sphere) are on the bounding meridian at latitude $\tan^{-1}(-\frac{4}{3})$ or $-53°\ 07'\ 48''\cdot368$.

The range of scale is small in the central part of the map, and Magis compared it favourably with several more commonly used projections.

34. Projection of a hemisphere within an equilateral triangle (Adams 1925). To map a hemisphere conformally within an equilateral triangle, we take

$$\text{sm}\,w = r, \qquad\qquad (34.1)$$

where r is a stereographic projection with $m = \frac{1}{2}$ at the origin. Various different aspects of the projection can be obtained by beginning with different aspects of the stereographic.

We have

$$\frac{dw}{dr} = \frac{1}{\text{cm}^2 w}, \qquad\qquad (34.2)$$

from which we can find the scale relative to the scale of the stereographic projection. The scale relative to the sphere is then

$$m = \tfrac{1}{2}(1 + p^2 + q^2).\,|1/\text{cm}^2 w|. \qquad\qquad (34.3)$$

Nature of the projection. Adams computed only the simplest case, with one pole at the origin, derived from the polar stereographic (16.3). The second hemisphere was in three pieces so that the whole sphere was mapped within a hexagon, as shown in Fig. 11.

Another case is illustrated in Fig. 12, with the north pole at one vertex and the south pole at the midpoint of the opposite side, derived from the transverse stereographic (15.2). Two equilateral triangles placed together map the sphere within a rhombus.

35. Projection of a hemisphere within a rhombus (Adams 1925).

Projection of a hemisphere and a half within an equilateral triangle. The projection of a hemisphere within an equilateral triangle (Fig. 11) can also be regarded as a projection of one-third of the sphere within a rhombus, the poles being located at the 120° angles and the range of longitude being 120°. We can map more or less than this range of longitude by first raising the stereographic coordinates to a power less than or greater than 1. Thus, to map a hemisphere within the rhombus, we need a range of longitude of 180° instead of 120°, and so the projection is given by

$$\text{sm}\,w = r^{2/3}. \qquad\qquad (35.1)$$

This will map a hemisphere and a half, 540° of longitude, within the unit circle, so that part of the map is repeated. There is

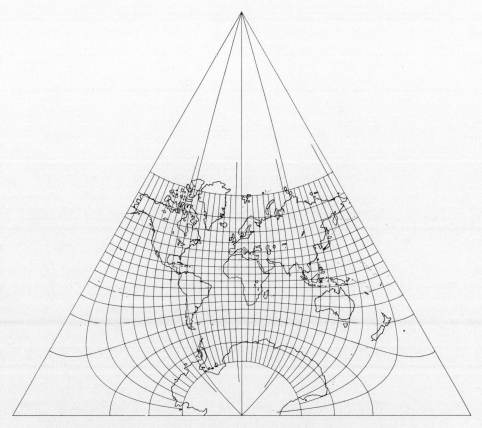

Fig. 10. Projection of the sphere within an equilateral triangle.

33. Projection of the sphere within an equilateral triangle (Cox 1935, Magis 1938).

Adams mentioned the possibility of mapping the sphere conformally within an equilateral triangle, but stated that "such a projection would not be especially important for geographic purposes, so that no computations were made for it." Coordinates of the projection were published by Magis 1938, but they are very inaccurate, no doubt because he followed Adams's complicated methods of computation.

The projection can be simply expressed by

$$\operatorname{sm} w = \tanh \tfrac{1}{4} \zeta, \tag{33.1}$$

(also noted by Gougenheim 1953), where $\tanh \tfrac{1}{4} \zeta$ is the Lagrange projection of the sphere within a circle, with origin on the equator and with $m = \tfrac{1}{4}$ at the origin. Thus, we first compute the coordinates of the Lagrange projection from (17.2), and then compute w as described in Sec. 32 above.

Scale. From (33.1) we have

$$\operatorname{cm}^2 w \, dw = \tfrac{1}{4} \operatorname{sech}^2 \tfrac{1}{4} \zeta \, d\zeta = \tfrac{1}{4}(1 - \operatorname{sm}^2 w) \, d\zeta, \tag{33.2}$$

whence

$$\frac{dw}{d\zeta} = \frac{1 - \operatorname{sm}^2 w}{4 \operatorname{cm}^2 w}. \tag{33.3}$$

There are four singular points, at each of the three vertices, and at the midpoint of one of the sides (in this case, at $w = -\frac{1}{2}K$).

The scale of the projection is given by $\sec\phi\,|dw/d\zeta|$, again with $m = \frac{1}{4}$ at the origin.

Nature of the projection. Fig. 10 illustrates a projection in which the north pole is at one vertex, and the south pole is at the midpoint of the opposite side, both poles being singular points. The other two singular points (two representations of the same point on the sphere) are on the bounding meridian at latitude $\tan^{-1}(-\frac{4}{3})$ or $-53°\ 07'\ 48''\cdot368$.

The range of scale is small in the central part of the map, and Magis compared it favourably with several more commonly used projections.

34. Projection of a hemisphere within an equilateral triangle (Adams 1925). To map a hemisphere conformally within an equilateral triangle, we take

$$\mathrm{sm}\,w = r, \tag{34.1}$$

where r is a stereographic projection with $m = \frac{1}{2}$ at the origin. Various different aspects of the projection can be obtained by beginning with different aspects of the stereographic.

We have

$$\frac{dw}{dr} = \frac{1}{\mathrm{cm}^2 w}, \tag{34.2}$$

from which we can find the scale relative to the scale of the stereographic projection. The scale relative to the sphere is then

$$m = \tfrac{1}{2}(1 + p^2 + q^2).\,|1/\mathrm{cm}^2 w|. \tag{34.3}$$

Nature of the projection. Adams computed only the simplest case, with one pole at the origin, derived from the polar stereographic (16.3). The second hemisphere was in three pieces so that the whole sphere was mapped within a hexagon, as shown in Fig. 11.

Another case is illustrated in Fig. 12, with the north pole at one vertex and the south pole at the midpoint of the opposite side, derived from the transverse stereographic (15.2). Two equilateral triangles placed together map the sphere within a rhombus.

35. Projection of a hemisphere within a rhombus (Adams 1925).

Projection of a hemisphere and a half within an equilateral triangle. The projection of a hemisphere within an equilateral triangle (Fig. 11) can also be regarded as a projection of one-third of the sphere within a rhombus, the poles being located at the 120° angles and the range of longitude being 120°. We can map more or less than this range of longitude by first raising the stereographic coordinates to a power less than or greater than 1. Thus, to map a hemisphere within the rhombus, we need a range of longitude of 180° instead of 120°, and so the projection is given by

$$\mathrm{sm}\,w = r^{2/3}. \tag{35.1}$$

This will map a hemisphere and a half, 540° of longitude, within the unit circle, so that part of the map is repeated. There is

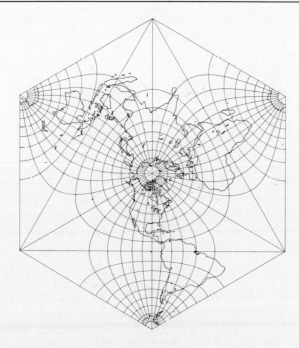

Fig. 11. Projection of a hemisphere within an equilateral triangle,
pole at the centre (sphere within a regular hexagon).

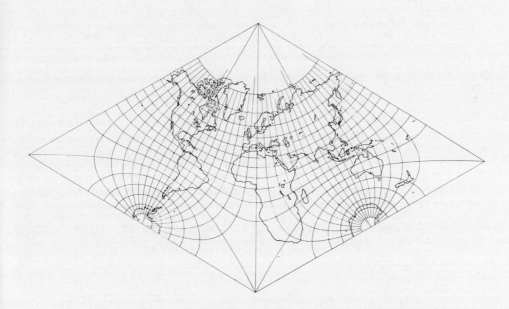

Fig. 12. Projection of a hemisphere within an equilateral triangle,
pole at a vertex (sphere within a rhombus).

now a singular point at the centre of the circle, and this remains a singular point at the centroid of the equilateral triangle, so that there is now a singular point at each vertex of the rhombus.

Differentiation of (35.1) gives

$$cm^2 \, w \, dw = \tfrac{2}{3} r^{-1/3} \, dr = \tfrac{2}{3} dr / \sqrt{(smw)}, \qquad (35.2)$$

whence

$$\frac{dw}{dr} = \frac{2}{3} \frac{1}{cm^2 w \sqrt{(smw)}} \qquad (35.3)$$

from which the scale can be found.

Projection of a hemisphere within a rhombus with poles at the 120° angles. If the origin is the pole, then from (16.3), $r = \tan \tfrac{1}{2}c \exp i\lambda$, so that

$$r^{2/3} = \tan^{2/3} \tfrac{1}{2} c (\cos \tfrac{2}{3}\lambda + i \sin \tfrac{2}{3}\lambda). \qquad (35.4)$$

After $s = r^{2/3}$ is computed, the rest of the computation is done in the same way as before (Sec. 32 above).

This leads to the simplest case of the projection, shown in Fig. 13, with the poles located at the 120° angles of the rhombus. Adams described two other cases.

Projection of a hemisphere within a rhombus with poles at the 60° angles. If we begin from a stereographic projection with the origin on the equator, and with the pole at the extremity of the central meridian (i.e. at the vertex of the equilateral triangle), r is computed from (15.2) and the transformation $smw = r^{2/3}$ produces a projection of a hemisphere within a rhombus with poles at the 60° angles, as shown in Fig. 14.

The computation can also be done, as Adams did it, by using the arc distance σ and azimuth A from the origin on the equator for forming the stereographic polar coordinates instead of the colatitude and longitude from the central meridian. That is, we compute

$$\cos \sigma = \cos \phi \cos \lambda, \qquad \tan A = \cot \phi \sin \lambda,$$

$$\tan \tfrac{1}{2}\sigma = \sqrt{[(1 - \cos \sigma)/(1 + \cos \sigma)]} \qquad (35.5)$$

and (σ, A) take the place of (c, λ) in (35.4).

Projection of a hemisphere within a rhombus with the pole at the intersection of the diagonals. A third version of the projection of a hemisphere within a rhombus can be derived by beginning from a stereographic projection with origin on the equator but with the pole located at an azimuth of 90° from the central meridian so that it is represented at the midpoint of the side of the equilateral triangle. This involves turning the stereographic projection through a right angle so that (q, p) of (15.2) become (p, q) of the new projection, or, if we use polar coordinates, so that the arc distance σ and the complement of the azimuth A from the origin on the equator (35.5) are used instead of the colatitude and longitude from the central meridian. The end result of the transformation $smw = r^{2/3}$ is then a projection of a hemisphere within a rhombus with the pole at the intersection of the diagonals, as shown in Fig. 15.

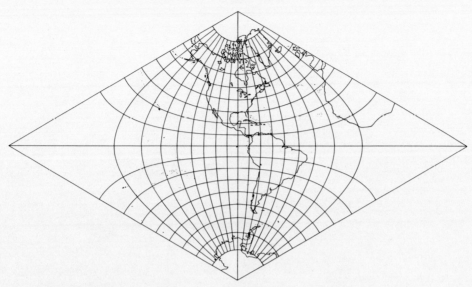

Fig. 13. Projection of a
hemisphere within a rhombus,
poles at the 120° angles.

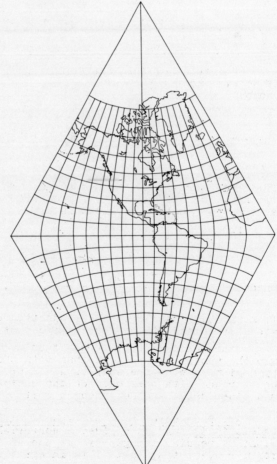

Fig. 14. Projection of a
hemisphere within a rhombus,
poles at the 60° angles.

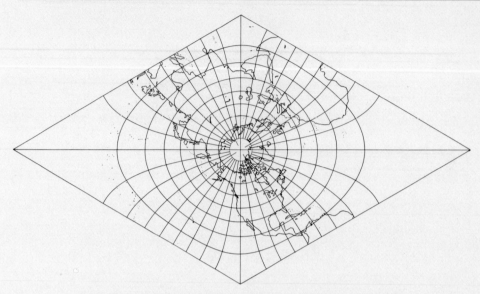

Fig. 15. Projection of a hemisphere within a rhombus,
pole at the intersection of the diagonals.

36. Projection of the sphere within a rhombus (Adams 1925).

Projection of three hemispheres within an equilateral triangle.
If we wish to fit 360° of longitude into the 120° angle of the
rhombus, we take

$$\text{sm } w = r^{1/3},\tag{36.1}$$

where r is a stereographic projection with origin at the pole, or

$$r^{1/3} = \tan^{1/3}\tfrac{1}{2}c\,(\cos\tfrac{1}{3}\lambda + i\sin\tfrac{1}{3}\lambda).\tag{36.2}$$

The hemisphere is now mapped three times within the equilateral
triangle, or the sphere is mapped within a rhombus with the poles
at the 120° angles, as shown in Fig. 16.

From (36.1) we have

$$\text{cm}^2 w\,\mathrm{d}w = \tfrac{1}{3}r^{-2/3}\,\mathrm{d}r = \tfrac{1}{3}\,\mathrm{d}r/\text{sm}^2 w,\tag{36.3}$$

whence
$$\frac{\mathrm{d}w}{\mathrm{d}r} = \frac{1}{3}\,\frac{1}{\text{sm}^2 w\,\text{cm}^2 w},\tag{36.4}$$

from which the scale of the projection can be found. There are
singular points at the four vertices of the rhombus.

Other cases of this projection are possible. Thus, we can
map the sphere within a rhombus with the poles at the 60° angles
by beginning with the stereographic projection with origin on the
equator and with the poles on the central meridian (15.2). This
and other cases are not thought worth illustrating.

37. Projection of a hemisphere within a regular hexagon
(Adams 1925). Adams derived a conformal projection of a hemisphere
within a regular hexagon by the use of Schwarz's integral (3.1)
for $n = 6$, stating that "This integral can be inverted in terms of

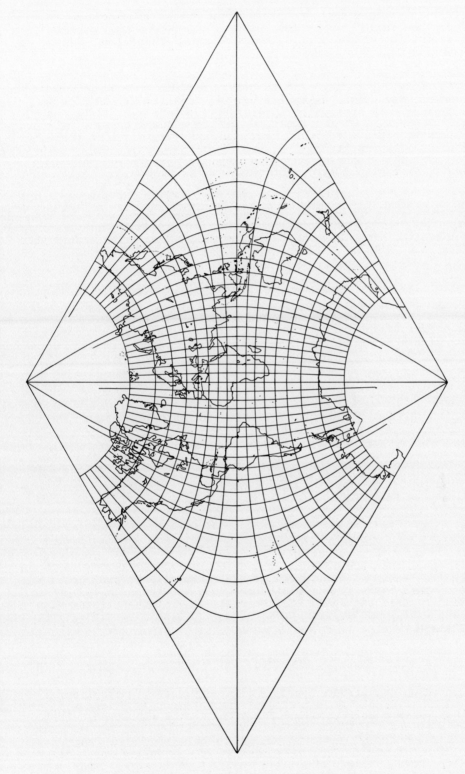

Fig. 16. Projection of the sphere within a rhombus, poles at the 120° angles.

elliptic functions, but the expressions are so complicated that it is more convenient to use the series for computation purposes." The definition of the projection in terms of Dixon functions is in fact not at all complicated.

Definition in terms of Dixon functions. We begin with the stereographic projection of the sphere divided into 12 spherical triangles, each with angles $\frac{1}{2}\pi$, $\frac{1}{2}\pi$, $\frac{1}{3}\pi$; that is, if the pole is the origin, each hemisphere is divided into six sectors by meridians 60° apart. Let the two poles be represented by 0, ∞, three vertices be represented by i, $i\omega$, and $i\omega^2$, and the other three vertices be represented by $-i$, $-i\omega$, and $-i\omega^2$. Then, from (28.5), each spherical triangle can be conformally represented on the infinite half-plane, with vertices at 0, 1, ∞, by

$$\xi = -\frac{4ir^3}{(i - r^3)^2}, \qquad 1 - \xi = \frac{(i + r^3)^2}{(i - r^3)^2}. \qquad (37.1)$$

The infinite half-plane can be conformally represented within the unit circle, with 0, 1, ∞ represented by 1, ω, ω^2 respectively, by

$$z = \frac{1 + \omega\xi}{1 + \omega^2\xi}, \qquad \xi = \frac{1 - z}{\omega^2 z - \omega}. \qquad (37.2)$$

Finally the unit circle can be conformally represented within an equilateral triangle, with 1, ω, ω^2 represented by 0, $K(1 - \omega)$, $K(1 - \omega^2)$ respectively, by

$$\mathrm{cm}\, w = z. \qquad (37.3)$$

By combining these transformations, the conformal projection of a hemisphere within a regular hexagon is given by

$$\mathrm{cm}\, w = \frac{1 - 2\sqrt{3}r^3 - r^6}{1 + 2\sqrt{3}r^3 - r^6}, \qquad \text{or} \quad \mathrm{sm}\, w = \frac{2^{2/3}3^{1/2}r(1 + r^6)^{2/3}}{1 + 2\sqrt{3}r^3 - r^6}. \qquad (37.4)$$

The origin is at the centre of the hexagon. The vertices of the hexagon are at $i\sqrt{3}K$, $i\omega\sqrt{3}K$, $i\omega^2\sqrt{3}K$, for which $\mathrm{cm}\, w = \omega$, and $-i\sqrt{3}K$, $-i\omega\sqrt{3}K$, $-i\omega^2\sqrt{3}K$, for which $\mathrm{cm}\, w = \omega^2$. At all vertices, $\mathrm{sm}\, w = 0$.

By differentiation of (37.4), we have

$$\mathrm{d}w/\mathrm{d}r = 2^{2/3}3^{1/2}(1 + r^6)^{-1/3}, \qquad (37.5)$$

which shows that the scale at the origin is $2^{-1/3}3^{1/2} = 1\cdot374\,730$ (since the scale at the origin of the stereographic projection is $\frac{1}{2}$). This makes it desirable to reduce the scale of the projection so that the scale at the origin is $\frac{1}{2}$ (thus agreeing with the scale of Adams's projection), and accordingly to define the projection by

$$\mathrm{cm}\,(2^{-2/3}3^{-1/2}w) = \frac{1 - 2\sqrt{3}r^3 - r^6}{1 + 2\sqrt{3}r^3 - r^6}. \qquad (37.6)$$

All coordinates derived from (37.4) are therefore multiplied by $2^{-2/3}3^{-1/2} = 0\cdot363\,708$. The vertex formerly at $i\sqrt{3}K$ is now at $2^{-2/3}iK$, with a corresponding reduction in the coordinates of the other vertices.

Computation of coordinates by series. A series for w can be developed from the expressions above, but it is more simply derived

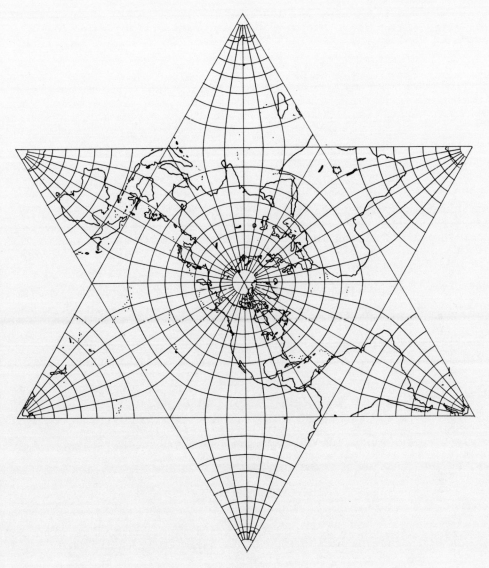

Fig. 17. Projection of a hemisphere within a regular hexagon,
(sphere within a six-pointed star).

from Schwarz's integral (3.1) in the form

$$w = \int_0^r [\, 1 - (ir)^6 \,]^{-1/3}\, dr$$

$$= r - \frac{1}{3} \cdot \frac{1}{7}\, r^7 + \frac{1.4}{3.6} \cdot \frac{1}{13}\, r^{13} - \frac{1.4.7}{3.6.9} \cdot \frac{1}{19}\, r^{19} + \cdots, \qquad (37.7)$$

or $\qquad w = \qquad\quad r \qquad\qquad -0\cdot004\,027\, r^{31}$

$\qquad\qquad\qquad -0\cdot047\,619\, r^7 \qquad +0\cdot002\,999\, r^{37}$

$\qquad\qquad\qquad +0\cdot017\,094\, r^{13} \qquad -0\cdot002\,335\, r^{43}$

$\qquad\qquad\qquad -0\cdot009\,097\, r^{19} \qquad +0\cdot001\,878\, r^{49}$

$\qquad\qquad\qquad +0\cdot005\,761\, r^{25} \qquad\quad -\cdots, \qquad\qquad (37.8)$

which is practicable for $|r| \not> 0\cdot 845$. In the series used by Adams, all terms are positive because of a different orientation. It can be noted that Adams's constant M is equal to $2^{-2/3}K$.

The inverse series is

$$r = w + \frac{1}{21} w^7 - \frac{1}{819} w^{13} + \frac{368}{9\ 80343} w^{19} - \frac{9692}{1029\ 36015} w^{25} + \cdots , \quad (37.9)$$

or

$$
\begin{aligned}
r = \quad & w && +0\cdot 000\ 029\, w^{31} \\
& +0\cdot 047\ 619\, w^7 && -0\cdot 000\ 010\, w^{37} \\
& -0\cdot 001\ 221\, w^{13} && +0\cdot 000\ 003\, w^{43} \\
& +0\cdot 000\ 375\, w^{19} && -0\cdot 000\ 001\, w^{49} \\
& -0\cdot 000\ 094\, w^{25} && + \cdots ,
\end{aligned}
\qquad (37.10)
$$

which can be used for $|w| \not> 1$, or very nearly the whole area of the hexagon.

For points beyond the range of the series, we use (37.4) and the method of Sec. 32, later reducing the coordinates as in (37.6).

Nature of the projection. The projection, called by Adams a rhombic projection of the world in a six-pointed star, is shown in Fig. 17, with the second hemisphere in six sectors placed around the hexagon. It can be seen from the symmetry that only one-twelfth of a hemisphere, or 30° of longitude, need be computed.

38. Projection of the sphere upon a regular tetrahedron (Lee 1965).

Derivation of formula. The surface of the sphere can be divided into four equiangular equilateral spherical triangles, each of which can be conformally represented upon an equilateral triangular face of the regular tetrahedron. Each such spherical triangle is composed of six smaller spherical triangles, each with angles $\frac{1}{2}\pi$, $\frac{1}{3}\pi$, $\frac{1}{3}\pi$. Each of these triangles can be represented upon the infinite half-plane, with origin at a vertex with angle $\frac{1}{3}\pi$, by the function ξ given in (24.8),

$$\xi = \frac{r^3(2\sqrt{2} + r^3)^3}{(1 + 5\sqrt{2}r^3 - r^6)^2}, \qquad 1 - \xi = \frac{(1 - 2\sqrt{2}r^3)^3}{(1 + 5\sqrt{2}r^3 - r^6)^2}. \quad (38.1)$$

By representing one such triangle within a sector of the unit circle with angle $\frac{1}{3}\pi$ by (31.2), and then representing the unit circle within an equilateral triangle by (32.1), we get

$$\text{sm}^3 w = (1 - \sqrt{1 - \xi})^2 / \xi. \quad (38.2)$$

Equations (38.1) and (38.2) define the conformal projection of the sphere upon a regular tetrahedron, but they can be expressed in a simpler form. Componendo and dividendo, we get

$$\frac{1 + \text{sm}^3 w}{1 - \text{sm}^3 w} = \frac{1}{\sqrt{1 - \xi}}. \quad (38.3)$$

By squaring both sides, and dividendo again, we get

$$\frac{4\,\text{sm}^3 w}{\text{cm}^6 w} = \frac{\xi}{1 - \xi}, \quad (38.4)$$

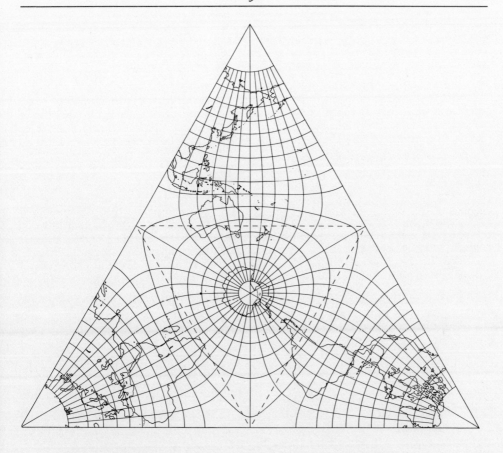

Fig. 18. Projection of the sphere upon a regular tetrahedron,
with the north pole at the common vertex of three faces,
and the south pole at the centre of the fourth face.

whence finally

$$\left(\frac{\xi}{1-\xi}\right)^{1/3} = \frac{r(2\sqrt{2}+r^3)}{1-2\sqrt{2}r^3} = \frac{2^{2/3} \operatorname{sm} w}{\operatorname{cm}^2 w}. \tag{38.5}$$

One solution to this equation is

$$r = 2^{1/6} \operatorname{sm} \tfrac{1}{2}w \operatorname{cm} \tfrac{1}{2}w, \tag{38.6}$$

which is probably the simplest form in which the projection can be
expressed. We can also derive

$$\operatorname{sm}^3 \tfrac{1}{2}w = \tfrac{1}{2}(1-\sqrt{1-2\sqrt{2}r^3}), \qquad \operatorname{cm}^3 \tfrac{1}{2}w = \tfrac{1}{2}(1+\sqrt{1-2\sqrt{2}r^3}). \tag{38.7}$$

Computation of coordinates by series. For the computation of
coordinates we can express r as a series in powers of w thus:

$$2^{5/6} r = 2 \operatorname{sm} \tfrac{1}{2}w \operatorname{cm} \tfrac{1}{2}w$$

$$= w - \frac{1}{16} w^4 + \frac{1}{448} w^7 - \frac{1}{14336} w^{10} + \frac{3}{14\,90944} w^{13} - \cdots, \tag{38.8}$$

which can be solved by an iterative method, beginning with $w = 2^{5/6}r$ as the first approximation. Within one triangular face, the series is rapidly convergent everywhere, even at a vertex, and the following terms are adequate to give 6-decimal accuracy:

$$
\begin{aligned}
w = \quad & 1 \cdot 781\ 797\,r & & -0 \cdot 001\ 477\,(w^{19} \times 10^{-6}) \\
& +0 \cdot 625 \quad (w^4 \times 10^{-1}) & & +0 \cdot 000\ 385\,(w^{22} \times 10^{-7}) \\
& -0 \cdot 223\ 214\,(w^7 \times 10^{-2}) & & -0 \cdot 000\ 099\,(w^{25} \times 10^{-8}) \\
& +0 \cdot 069\ 754\,(w^{10} \times 10^{-3}) & & +0 \cdot 000\ 025\,(w^{28} \times 10^{-9}) \\
& -0 \cdot 020\ 121\,(w^{13} \times 10^{-4}) & & - \ \cdots. \\
& +0 \cdot 005\ 539\,(w^{16} \times 10^{-5}) & & \hspace{3cm} (38.9)
\end{aligned}
$$

Although formally equation (38.6) applies to the projection of the whole sphere, the series (38.9) is not necessarily convergent beyond the first triangular face. It is therefore necessary to compute each face separately and to refer the coordinates to a common origin later.

The series (38.9) can be reversed to give the direct formula, where $R = 2^{5/6}r$,

$$
w = R + \frac{1}{16}R^4 + \frac{3}{224}R^7 + \frac{1}{256}R^{10} + \frac{35}{26624}R^{13} + \cdots , \qquad (38.10)
$$

or
$$
\begin{aligned}
w = \quad & R & & +0 \cdot 000\ 186\,R^{19} \\
& +0 \cdot 062\ 5 \quad R^4 & & +0 \cdot 000\ 074\,R^{22} \\
& +0 \cdot 013\ 393\,R^7 & & +0 \cdot 000\ 031\,R^{25} \\
& +0 \cdot 003\ 906\,R^{10} & & +0 \cdot 000\ 013\,R^{28} \\
& +0 \cdot 001\ 315\,R^{13} & & + \ \cdots. \\
& +0 \cdot 000\ 481\,R^{16} & & \hspace{3cm} (38.11)
\end{aligned}
$$

This series is useful when $|R| \not> 0 \cdot 86$, or within about 50° of the origin, but is slowly convergent elsewhere. In series (38.9) the convergence of the iterations may be slow in the vicinity of a vertex, and it can be accelerated by using series (38.11) to get a first approximation.

Scale. By differentiation of (38.6) we get

$$
\frac{dw}{dr} = \frac{2^{5/6}}{\mathrm{cm}^3\,\tfrac{1}{2}w - \mathrm{sm}^3\,\tfrac{1}{2}w} = \frac{2^{5/6}}{\sqrt{1 - 2\sqrt{2}r^3}}, \qquad (38.12)
$$

which shows that the only singular points are those where $\mathrm{cm}^3\,\tfrac{1}{2}w = \mathrm{sm}^3\,\tfrac{1}{2}w$, that is, at the vertices of the triangular face.

The scale is given by

$$
m = \tfrac{1}{2}(1 + p^2 + q^2) \cdot |dw/dr|, \qquad (38.13)
$$

which in conjunction with (38.12) can be used for evaluating the scale in certain simple cases. Thus, on the x-axis ($v = 0$, $q = 0$), the scale is given by

$$
m = \frac{2^{-1/6} + 2^{1/6}\,\mathrm{sm}^2\,\tfrac{1}{2}u\,\mathrm{cm}^2\,\tfrac{1}{2}u}{\mathrm{cm}^3\,\tfrac{1}{2}u - \mathrm{sm}^3\,\tfrac{1}{2}u} = \frac{2^{-1/6}(1 + p^2)}{\sqrt{1 - 2\sqrt{2}p^3}}. \qquad (38.14)
$$

To evaluate the scale in the general case, we have from (38.8),

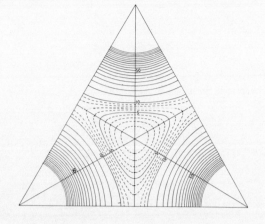

Fig. 19. Isomegeths (curves of
constant scale) in one triangular face of the projection
of the sphere upon a regular
tetrahedron.

$$\frac{\mathrm{d}r}{\mathrm{d}w} = 2^{-5/6}\left(1 - \frac{1}{4}w^3 + \frac{1}{64}w^6 - \frac{5}{7168}w^9 + \frac{3}{1\,14688}w^{12} - \cdots\right), \qquad (38.15)$$

or $\quad \dfrac{\mathrm{d}r}{\mathrm{d}w} = 0\cdot561\,231 \qquad\qquad +0\cdot015\,754\,(w^{18} \times 10^{-6})$

$\qquad\qquad -1\cdot403\,078\,(w^3 \times 10^{-1}) \qquad -0\cdot004\,756\,(w^{21} \times 10^{-7})$

$\qquad\qquad +0\cdot876\,923\,(w^6 \times 10^{-2}) \qquad0\cdot001\,385\,(w^{24} \times 10^{-8})$

$\qquad\qquad -0\cdot391\,484\,(w^9 \times 10^{-3}) \qquad -0\cdot000\,392\,(w^{27} \times 10^{-9})$

$\qquad\qquad +0\cdot146\,806\,(w^{12} \times 10^{-4}) \qquad +\ \cdots.$

$\qquad\qquad -0\cdot049\,742\,(w^{15} \times 10^{-5}) \hfill (38.16)$

The scale is now given by

$$m = \frac{\frac{1}{2}(1 + p^2 + q^2)}{|\mathrm{d}r/\mathrm{d}w|}. \qquad (38.17)$$

A formula giving the scale directly in terms of the coordinates
has also been developed, but it is too lengthy for any journal
to print.

At the centroid of each face, the scale is $2^{-1/6} = 0\cdot890\,899$;
at the midpoint of each edge it is $2^{1/3}3^{-1/4} = 0\cdot957\,333$. At the
midpoint of each edge, the scale relative to the scale at the centroid is therefore $2^{1/2}3^{-1/4} = 1\cdot074\,570$, or the scale enlargement
is $7\cdot4570\%$. Isomegeths for values less than this are closed
curves around the centroid; isomegeths for values greater are
closed curves around the vertex. The scale is very nearly constant,
at about $7\frac{1}{2}\%$, along the straight lines joining the midpoints of
the edges. Fig. 19 shows the pattern of isomegeths in one face of
the tetrahedron; the numerical values are percentage increases of
scale over the scale at the centroid.

Nature of the projection. The conformal projection of the
sphere upon a regular tetrahedron represents the sphere within
four equal equilateral triangles, each of which is a face of the
tetrahedron and each of which represents an equilateral spherical
triangle of which each angle is 120° and each side is $\cos^{-1}(-\frac{1}{3})$
or 109° 28′ 16″·394. The centroid of each face is the antipodes
of the common vertex of the three adjoining faces. The circum-

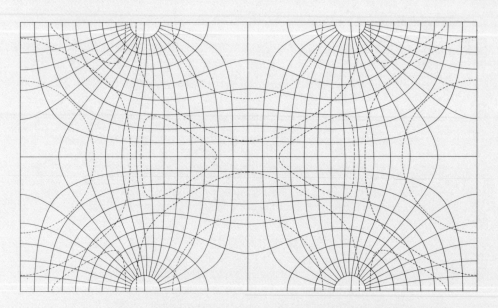

Fig. 20. Projection of the sphere upon a regular tetrahedron,
with the equator as the common edge of two faces and a
median of each of the other two.

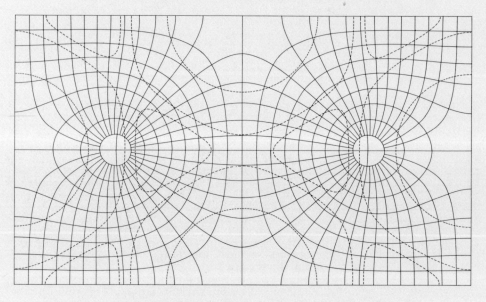

Fig. 21. Transverse aspect of the projection shown in Fig. 20.
The two graticules are identical.

Fig. 22. Repetition of the sphere on conformal tetrahedric projection.

radius of each spherical triangle is the supplement of the side (70° 31' 43"·606) and the inradius is half the side (54° 44' 08"·197). The faces can be assembled to form one larger equilateral triangle or a rhombus.

Fig. 18 illustrates a conformal tetrahedric projection of the sphere in which one vertex is placed at the north pole and the other three at points in the sea on the parallel of south latitude $\sin^{-1}\frac{1}{3}$, at longitudes 20° W, 100° E, and 140° W. The land masses are clear of the singular points and are represented without undue distortion. Only one-sixth of the polar face and half of the other face need be computed for this graticule. The latter face has an origin given by $\sin \phi_0 = \frac{1}{3}$, or $\phi_0 = 19° 28' 16"·394$.

If one face is cut along a median, the parts can be assembled to represent the sphere within a rectangle with sides in the ratio $4:\sqrt{3}$, or, if two faces are cut along medians, a rectangle with sides in the ratio $\sqrt{3}:1$. Fig. 20 illustrates an example of the latter case, the equator being represented by the common edge of two faces and by a median of each of the other two, so that there are two singular points on the equator and two on a meridian. The scale is nearly constant along the diagonals of the rectangle. Fig. 20 shows also the 5%, 10%, and 50% isomegeths. Only two half-faces need to be computed for this graticule; one has an origin on the central meridian at a latitude given by $\tan \phi_0 = \sqrt{2}$, or $\phi_0 = 54° 44' 08"·197$, the other has an origin on the equator at a longitude of 54° 44' 08"·197 reckoned from the edge of the face.

The great circle which is the common edge of two faces is at right angles to the great circle which is the common edge of the other two faces. If one great circle is the equator, the other will be a meridian. This leads to a curious property of the projection of Fig. 20 (possessed also by projections of the sphere upon a cube and upon a regular octahedron), that the graticule of the transverse aspect is identical with that of the direct aspect. Thus, if Fig. 20 is the direct aspect, the graticule of the transverse aspect shown in Fig. 21 is obtained merely by rearranging the pieces.

The projection upon a tetrahedron possesses one advantage over the projections upon the other regular polyhedra: the faces can be assembled so that every face matches all adjoining faces, and the entire sphere is represented without gaps or overlaps. This makes it possible to repeat the sphere an indefinite number of times, without discontinuities other than those at the singular points. Fig. 22 shows the sphere repeated twelve times.

The projection can be remarkably useful for groupings of continents. Fig. 23 shows an example (one of a number of suggestions by H. J. Andrews) in which Europe, Africa, Asia and Australasia are represented, using the same graticule as in Fig. 19 but with the meridians renumbered so that land areas are moved 20° eastward. The 5%, 10%, and 50% isomegeths are shown. The same map could be extended to include North America and the north Pacific Ocean, and Antarctica.

The conformal tetrahedric projection is admirably suited to the representation of roughly triangular areas of continental size. An example including Africa and the Arabian peninsula is shown in Fig. 24. This has an origin at latitude 8° N, longitude 21° E, and an overall scale constant applied to give a range of scale enlargement from $+3\frac{1}{2}$% to $-3\frac{1}{2}$% over most of the mainland of Africa.

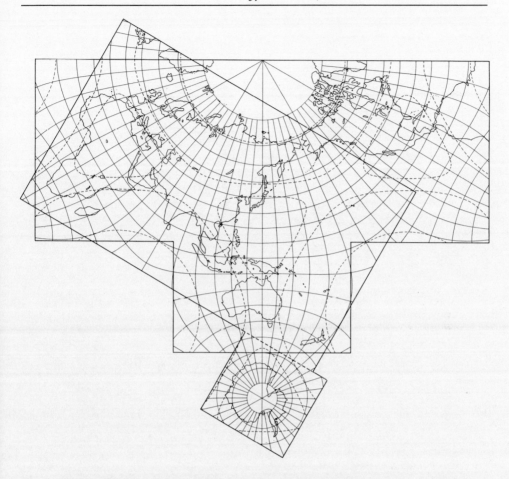

Fig. 23. Continental groupings on conformal tetrahedric projection.

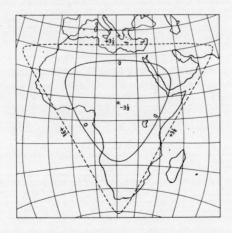

Fig. 24. Africa on conformal tetrahedric projection.

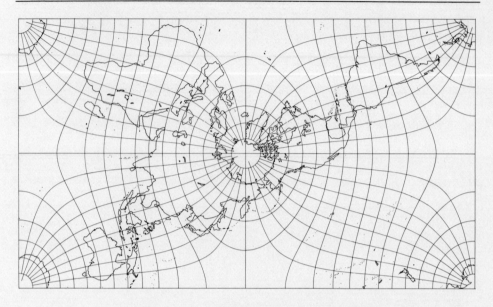

Fig. 25. Projection of the sphere upon a regular tetrahedron,
with the poles at midpoints of edges.

The foregoing description of the conformal projection of the
sphere upon a regular tetrahedron is substantially that published
by Lee 1973. The remarks concerning Figs. 20 and 21 assume that
the graticule upon one face can be the transverse aspect of the
graticule upon another face, or, alternatively, we can regard the
two figures as projections of the sphere within a rectangle and
ignore their derivation from the faces of a tetrahedron. If the
sphere is repeated indefinitely, Fig. 21 can be derived from Fig.
20 merely by moving the boundary of the rectangle.

Mr Thomas Wray, in personal correspondence, contended that
the projection upon a tetrahedron, with all faces together, should
be considered as an entity, and that one particular graticule
should be nominated as the direct aspect. He selected the case
where each pole is at the midpoint of an edge, so that all four
faces have identical graticules. The centroids are then at lat-
itudes given by $\cot \phi_0 = \pm\sqrt{2}$, or $\phi_0 = \pm35° \ 15' \ 51''\cdot803$, and the
singular points are in the same latitudes, at 90° longitude from
the centroids. This case is illustrated in Fig. 25, again representing the sphere within a rectangle with sides in the ratio $\sqrt{3}:1$,
and with the singular points at the midpoints of the sides. If
we accept Wray's choice of Fig. 25 as the direct aspect, then
Figs. 20 and 21 could be regarded as transverse aspects. However,
Wray considers that, because the straight lines joining the midpoints of opposite edges of a regular tetrahedron form a triad of
mutually perpendicular lines, and one of these lines, that joining
the two poles, has been selected as defining the direct aspect,
the other two lines should define the first and second transverse
aspects — which results in identical graticules for all three
aspects. Figs. 20 and 21 are then regarded as transverse equi-
oblique.

The fact that the projection upon a tetrahedron fits so
neatly into a rectangle suggested that Jacobian elliptic functions could deal with it. From a preliminary sketch of Fig. 25,

Wray recognized the appropriate function. With origin at a geographic pole, the projection is given by

$$\text{sd}\, w \;=\; 2\exp\zeta, \tag{38.18}$$

where the modulus is $\sin 15°$, and the scale differs only slightly from that given by the definition in (38.6). The scale of (38.18) is greater than the scale of (38.6) in the ratio of the real period of the Jacobian functions to $\tfrac{1}{2}\sqrt{3}$ times the real period of the Dixon functions, or $2^{-1/3}\,3^{1/4} = 1\cdot044\ 569$. In (38.18), $m = 1$ at the origin (the pole), and the minimum scale is $m = 2^{-1/2}\,3^{1/4} = 0\cdot930\ 065$ at the centroid of a triangular face.

The similarity of the graticule to that of the Peirce projection (Fig. 28) was also noticed by Wray. Equations (38.18) and (43.1) differ only in the modulus and constants.

By locating the origin in other positions, Wray has shown that the projection can be defined by other Jacobian functions, or by the Weierstrass elliptic function, all without using the stereographic as an intermediate transformation. The conformal projection of the sphere upon a regular tetrahedron can therefore be expressed more simply in terms of Jacobian functions than Dixon functions, and would have been more appropriately treated in the next chapter.

39. Projection of the sphere upon a regular octahedron (Adams 1928).

Derivation of formula. The area of the sphere which is represented by one of the eight faces of the regular octahedron is a trirectangular spherical triangle composed of six spherical triangles, each with angles $\tfrac{1}{2}\pi$, $\tfrac{1}{3}\pi$, $\tfrac{1}{4}\pi$. Each such spherical triangle can be conformally represented upon the infinite half-plane, with origin at the vertex with angle $\tfrac{1}{3}\pi$, by the function ξ given in (25.9), i.e.

$$\left.\begin{aligned}
\xi &= \frac{4r^3\,(2\sqrt{2} + 7r^3 - 2\sqrt{2}r^6)^3}{(1 + 22\sqrt{2}r^3 + 22\sqrt{2}r^9 - r^{12})^2}, \\[2mm]
1 - \xi &= \frac{(1 - 5\sqrt{2}r^3 - r^6)^4}{(1 + 22\sqrt{2}r^3 + 22\sqrt{2}r^9 - r^{12})^2}.
\end{aligned}\right\} \tag{39.1}$$

The infinite half-plane can be represented within a sector of the unit circle with angle $\tfrac{1}{3}\pi$ by (31.2), so that, from (32.1),

$$\text{sm}^3 w \;=\; (1 - \sqrt{1 - \xi})^2/\xi, \tag{39.2}$$

leading to

$$\text{sm}\, w \;=\; \frac{r(2\sqrt{2} - r^3)}{1 + 2\sqrt{2}r^3}, \tag{39.3}$$

which gives the conformal representation of the trirectangular spherical triangle within the equilateral triangular face of the octahedron. Equation (39.3) is tantalizingly similar to (38.5), but a similar simple solution for r has not been obtained.

Computation of coordinates by series. In the general case, coordinates can be computed by the usual procedure of separating the real and the imaginary parts of w from the numerical value of $\text{sm}\, w$ described in Sec. 32.

For points close to the origin, the coordinates can be expressed as a series in powers of R, where

$$R = 2\sqrt{2}r = 2 \cdot 828\ 427\ r. \tag{39.4}$$

We then have

$$sm\,w = R - \frac{9}{2^6}R^4\left(1 + \frac{1}{2^3}R^3\right)^{-1}$$

$$= R - \frac{9}{2^6}R^4 + \frac{9}{2^9}R^7 - \frac{9}{2^{12}}R^{10} + \frac{9}{2^{15}}R^{13} - \cdots. \tag{39.5}$$

Substituting this series and its powers in

$$w = sm\,w + \frac{2}{3}\cdot\frac{1}{4}\,sm^4w + \frac{2.5}{3.6}\cdot\frac{1}{7}\,sm^7w + \frac{2.5.8}{3.6.9}\cdot\frac{1}{10}\,sm^{10}w + \cdots, \tag{39.6}$$

we thus obtain

$$w = R + \frac{5}{192}R^4 + \frac{103}{32256}R^7 + \frac{23}{41472}R^{10} + \frac{46927}{41\,40\,56448}R^{13} + \cdots, \tag{39.7}$$

or

$$
\begin{aligned}
w = \quad & R & &+0\cdot000\ 025\ R^{16} \\
&+0\cdot026\ 042\ R^4 & &+0\cdot000\ 006\ R^{19} \\
&+0\cdot003\ 193\ R^7 & &+0\cdot000\ 002\ R^{22} \\
&+0\cdot000\ 555\ R^{10} & &+\cdots. \\
&+0\cdot000\ 113\ R^{13} & &
\end{aligned}
\tag{39.8}
$$

This series is suitable for use when $|R| \not> 0\cdot9.$ or within about 40° of the centre of the trirectangular triangle, but is too slowly convergent beyond that distance. The inverse series is

$$2\sqrt{2}r = w - \frac{5}{192}w^4 - \frac{31}{64512}w^7 - \frac{2111}{743\ 17824}w^{10} - \frac{2\ 55727}{9\ 27486\ 44352}w^{13} - \cdots, \tag{39.9}$$

which is also too slowly convergent to be useful beyond 40° from the centre.

Scale. From (39.3) we can derive

$$cm\,w = \frac{(1 - 5\sqrt{2}r^3 - r^6)^{2/3}}{1 + 2\sqrt{2}r^3}, \tag{39.10}$$

and also

$$cm^2\,w\,dw = \frac{2\sqrt{2}(1 - 5\sqrt{2}r^3 - r^6)}{(1 + 2\sqrt{2}r^3)^2}\,dr, \tag{39.11}$$

whence

$$\frac{dw}{dr} = \frac{2\sqrt{2}}{(1 - 5\sqrt{2}r^3 - r^6)^{1/3}}, \tag{39.12}$$

from which the scale can be found. The scale at the origin is $\sqrt{2}$, so that an overall scale reduction should be applied to the projection. This scale reduction can be applied at the beginning, by replacing $sm\,w$ in (39.3) by $sm\sqrt{2}w$, to give a scale of 1 at the origin, but the improvement in the series is slight.

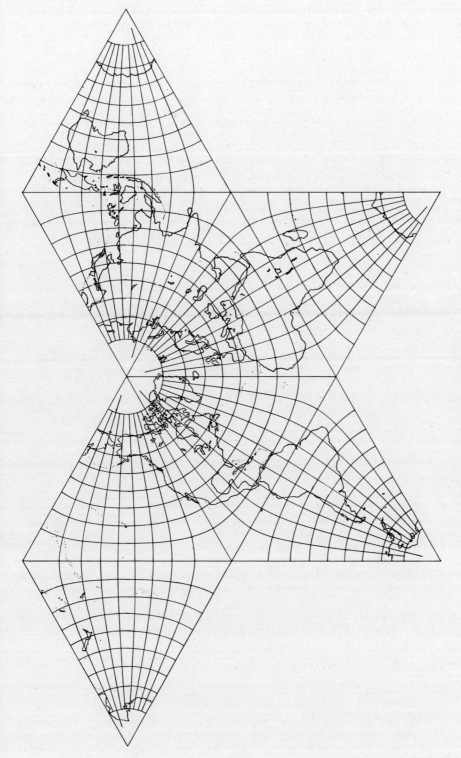

Fig. 26. Projection of the sphere upon a regular octahedron.

Nature of the projection. The eight equilateral triangular faces of the octahedron have been customarily arranged in the formation shown in Fig. 26, which has been styled a "butterfly map". Each face has a pole at one vertex, with a quadrant of the equator represented by the opposite side. The latitude of the origin is given by cot ϕ_0 = √2, or ϕ_0 = 35° 15' 51"·803.

If the equator is turned through a right angle to take up a position formerly occupied by a meridian, the graticule on every face will be rotated so that the pole moves from one vertex to another, and the new graticule is identical with the original one. This is not, strictly speaking, the transverse aspect, in which the equator would be a median of a face, and the pole would be the midpoint of an edge of an adjoining face.

The "butterfly map" has been popularized by Cahill, but he attributed the theory and computations to Adams. We have not seen Adams's theory and computations, although it is known that type-written copies have been deposited in certain libraries.

40. Projection of the sphere upon a regular icosahedron (Lee 1971).

Derivation of formula. Each of the twenty faces of the regular icosahedron represents a spherical triangle composed of six smaller spherical triangles, each with angles $\frac{1}{2}\pi$, $\frac{1}{3}\pi$, $\frac{1}{5}\pi$. Each such spherical triangle is conformally represented upon the infinite half-plane, with origin at a vertex with angle $\frac{1}{3}\pi$, by the function ξ in (26.10), i.e.

$$\xi = \frac{25\sqrt{5}r^3(8 + 57\sqrt{5}r^3 - 228r^6 + 494\sqrt{5}r^9 + 228r^{12} + 57\sqrt{5}r^{15} - 8r^{18})^3}{(8 + 580\sqrt{5}r^3 + 3915r^6 + 20010\sqrt{5}r^9 - 190095r^{12} - 190095r^{18}}$$
$$- 20010\sqrt{5}r^{21} + 3915r^{24} - 580\sqrt{5}r^{27} + 8r^{30})^2 ,$$

$$1 - \xi = \frac{64(1 - 11\sqrt{5}r^3 - 33r^6 + 11\sqrt{5}r^9 + r^{12})^5}{(8 + 580\sqrt{5}r^3 + 3915r^6 + 20010\sqrt{5}r^9 - 190095r^{12} - 190095r^{18}}$$
$$- 20010\sqrt{5}r^{21} + 3915r^{24} - 580\sqrt{5}r^{27} + 8r^{30})^2 . \qquad (40.1)$$

By the same procedure as that used for the tetrahedron (Sec. 38), we then obtain

$$\frac{2^{2/3} \operatorname{sm} w}{\operatorname{cm}^2 w} = \left(\frac{\xi}{1 - \xi}\right)^{1/3}, \qquad (40.2)$$

or

$$\frac{\operatorname{sm} w}{\operatorname{cm}^2 w} = 2^{-8/3}5^{5/6} \frac{r(8 + 57\sqrt{5}r^3 - 228r^6 + 494\sqrt{5}r^9 + 228r^{12} + 57\sqrt{5}r^{15} - 8r^{18})}{(1 - 11\sqrt{5}r^3 - 33r^6 + 11\sqrt{5}r^9 + r^{12})^{5/3}},$$
$$\qquad (40.3)$$

which is probably the most compact way of expressing the conformal projection of the sphere upon a regular icosahedron, although it is not the most suitable form for computation.

Computation of coordinates by series. If we let

$$R^3 = 50\sqrt{5}r^3, \qquad \text{or} \qquad R = 2^{1/3}5^{5/6}r = 4\cdot817\,462\,r, \qquad (40.4)$$

we can express (40.3) as

$$\frac{\text{sm } w}{\text{cm}^2 w} = \frac{R\left(1 + \dfrac{57}{2^4 5^2}R^3 - \dfrac{57}{2^3 5^5}R^6 + \dfrac{247}{2^5 5^7}R^9 + \dfrac{57}{2^5 5^{10}}R^{12} + \dfrac{57}{2^8 5^{12}}R^{15} - \dfrac{1}{2^6 5^{15}}R^{18}\right)}{\left(1 - \dfrac{11}{2.5^2}R^3 - \dfrac{33}{2^2 5^5}R^6 + \dfrac{11}{2^3 5^7}R^9 + \dfrac{1}{2^4 5^{10}}R^{12}\right)^{5/3}}$$

$$= R + \frac{611}{2^4.3.5^2}R^4 + \frac{1\,45733}{2^5.3^2.5^5}R^7 + \frac{189\,09029}{2^6.3^4.5^7}R^{10} + \cdots. \tag{40.5}$$

From this series we then obtain

$$w = \frac{\text{sm } w}{\text{cm}^2 w} - \frac{1}{2}\left(\frac{\text{sm } w}{\text{cm}^2 w}\right)^4 + \frac{6}{7}\left(\frac{\text{sm } w}{\text{cm}^2 w}\right)^7 - 2\left(\frac{\text{sm } w}{\text{cm}^2 w}\right)^{10} + \cdots$$

$$= R + \frac{11}{1\,200}R^4 + \frac{4631}{63\,00000}R^7 + \frac{1\,38941}{16200\,00000}R^{10} + \cdots, \tag{40.6}$$

or $w =$

$$\begin{aligned}
&R & &+0\cdot189\,112\,(R^{16} \times 10^{-5}) \\
&+0\cdot091\,667\,(R^4 \times 10^{-1}) & &+0\cdot316\,991\,(R^{19} \times 10^{-6}) \\
&+0\cdot073\,508\,(R^7 \times 10^{-2}) & &+0\cdot557\,438\,(R^{22} \times 10^{-7}) \\
&+0\cdot085\,766\,(R^{10} \times 10^{-3}) & &+\cdots. \\
&+0\cdot120\,813\,(R^{13} \times 10^{-4})
\end{aligned} \tag{40.7}$$

This series is useful when $|R| \not> 1\cdot06$, or within about 25° of the origin.

The inverse series is

$$R = w - \frac{11}{1200}w^4 - \frac{5027}{126\,00000}w^7 - \frac{25\,93789}{9\,07200\,00000}w^{10} - \cdots, \tag{40.8}$$

or $R =$

$$\begin{aligned}
&w & &-0\cdot031\,122\,(w^{16} \times 10^{-5}) \\
&-0\cdot091\,667\,(w^4 \times 10^{-1}) & &-0\cdot037\,963\,(w^{19} \times 10^{-6}) \\
&-0\cdot039\,897\,(w^7 \times 10^{-2}) & &-0\cdot049\,217\,(w^{22} \times 10^{-7}) \\
&-0\cdot028\,591\,(w^{10} \times 10^{-3}) & &-\cdots, \\
&-0\cdot027\,893\,(w^{13} \times 10^{-4})
\end{aligned} \tag{40.9}$$

which can be used for $|w| \not> 1\cdot185$.

Computation of coordinates beyond the range of the series. To compute the coordinates of points further from the origin, we return to (40.1) and compute

$$\xi = 4R^3 P^3 / Q^2, \tag{40.10}$$

where

$$\begin{aligned}
P &= 1 & Q &= 1 \\
&+1\cdot425\,(R^3 \times 10^{-1}) & &+14\cdot5\,(R^3 \times 10^{-1}) \\
&-0\cdot228\,(R^6 \times 10^{-2}) & &+3\cdot915\,(R^6 \times 10^{-2}) \\
&+0\cdot098\,8\,(R^9 \times 10^{-3}) & &+4\cdot002\,(R^9 \times 10^{-3}) \\
&+0\cdot001\,824\,(R^{12} \times 10^{-4}) & &-1\cdot520\,76\,(R^{12} \times 10^{-4}) \\
&+0\cdot000\,091\,(R^{15} \times 10^{-5}), & &-0\cdot012\,166\,(R^{18} \times 10^{-6}) \\
& & &-0\cdot000\,256\,(R^{21} \times 10^{-7}) \\
& & &+0\cdot000\,002\,(R^{24} \times 10^{-8}).\,(40.11)
\end{aligned}$$

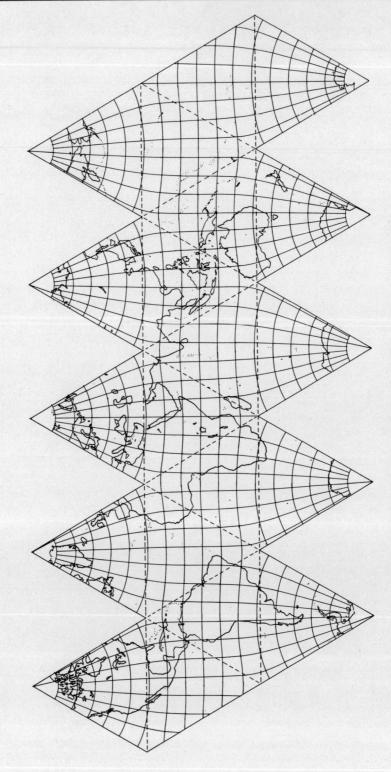

Fig. 27. Projection of the sphere upon a regular icosahedron.

Note that the series for Q has no term in R^{15}. A check can be applied from

$$1 - \xi = T^5/Q^2, \qquad (40.12)$$

where
$$\begin{aligned} T = \quad & 1 \\ & -2\cdot2\ (R^3 \times 10^{-1}) \\ & -0\cdot264\ (R^6 \times 10^{-2}) \\ & +0\cdot0176\ (R^9 \times 10^{-3}) \\ & +0\cdot000\,064\ (R^{12} \times 10^{-4}). \end{aligned} \qquad (40.13)$$

Then

$$\text{sm } w = (1 - \sqrt{1 - \xi})^2/\xi, \qquad (40.14)$$

and the value of w can be found by the procedure described in Sec. 32.

   Scale. The defining formula of this projection is so complicated that a simple expression for the scale is unlikely to be obtained. By differentiation of the series (40.6) we obtain

$$\frac{dw}{dr} = 2^{1/3} 5^{5/6} \left(1 + \frac{11}{300} R^3 + \frac{4631}{9\,00000} R^6 + \frac{1\,38941}{1620\,00000} R^9 + \cdots \right), \qquad (40.15)$$

from which the scale can be found. The scale at the origin is $2^{-2/3} 5^{5/6} = 2\cdot408\,731$, so that an overall scale reduction should be applied to the projection.

   Nature of the projection. Fig. 27 shows an example of the conformal projection of the sphere upon a regular icosahedron, in which the north pole is at a vertex in five of the faces, the south pole is at a vertex in another five faces, and the remaining ten faces each contain 36° of the equator. The north polar faces have an origin at latitude given by $\cot \phi_0 = 3 - \sqrt{5}$, or $\phi_0 = 52°\ 37'\ 21''\cdot475$, and the equatorial faces have an origin at $\cot \phi_0 = 3 + \sqrt{5}$, or $\phi_0 = 10°\ 48'\ 44''\cdot341$. In the table of coordinates, the two faces have been combined, with an origin at the vertex remote from the pole. A longitude interval of 2° is given so that meridians at 10° intervals can be drawn on any face.

   41. Alternative method of deriving series for projections within equilateral triangles. An alternative method of deriving the series for the projection of an equilateral spherical triangle within an equilateral plane triangle is provided by the transformation given in Kober's *Dictionary of Conformal Representations*, §14.3, p. 201. An equiangular equilateral circular triangle with angles $2\pi.\beta$ is represented within the unit circle s by

$$\frac{\Gamma(\frac{4}{3})\Gamma(\frac{1}{6} + \beta)}{\Gamma(\frac{2}{3})\Gamma(\frac{5}{6} + \beta)}\, w' = s\, \frac{F(\frac{5}{6} - \beta,\ \frac{1}{2} - \beta;\ \frac{4}{3};\ s^3)}{F(\frac{1}{6} - \beta,\ \frac{1}{2} - \beta;\ \frac{2}{3};\ s^3)}, \qquad (41.1)$$

where w' denotes the circular triangle reduced to unit circumradius and $F(a,\ b;\ c;\ x)$ denotes the hypergeometric series,

$$1 + \frac{ab}{c} x + \frac{1}{2!} \frac{a(a+1)b(b+1)}{c(c+1)} x^2 + \frac{1}{3!} \frac{a(a+1)(a+2)b(b+1)(b+2)}{c(c+1)(c+2)} x^3 + \cdots (41.2)$$

   Thus, to map the unit circle s within a plane equilateral triangle of circumradius K, we take $\beta = \frac{1}{6}$, $w' = w/K$, and get

$$w = z + \frac{1}{6}z^4 + \frac{5}{63}z^7 + \frac{4}{81}z^{10} + \cdots, \qquad (41.3)$$

which is the transformation $w = \mathrm{sm}^{-1}z$.

For the tetrahedron, octahedron, and icosahedron, we take $\beta = \frac{1}{3}, \frac{1}{4}, \frac{1}{5}$ respectively.

Tetrahedron. For the tetrahedron, we take $\beta = \frac{1}{3}$, $w' = \sqrt{2}r$, and we get

$$2^{1/3}(2^{1/2}r) = z + \frac{5}{48}z^4 + \frac{23}{576}z^7 + \frac{3593}{1\,65888}z^{10} + \cdots. \qquad (41.4)$$

If we let $2^{5/6}r = R$, reversal of this series gives

$$z = R - \frac{5}{48}R^4 + \frac{1}{288}R^7 - \frac{1}{1296}R^{10} + \cdots. \qquad (41.5)$$

From $w = \mathrm{sm}^{-1}z$, we now get

$$w = R + \frac{1}{16}R^4 + \frac{3}{224}R^7 + \frac{1}{256}R^{10} + \cdots, \qquad (41.6)$$

which is the series (38.10), previously derived by reversal of the series for $R = 2\,\mathrm{sm}\frac{1}{2}w\,\mathrm{cm}\frac{1}{2}w$.

Octahedron. For the octahedron, we take $\beta = \frac{1}{4}$, $w' = r\sqrt{2}(\sqrt{3}-1)$, and we get

$$2\sqrt{2}r = z + \frac{9}{2^6}z^4 + \frac{63}{2^{10}}z^7 + \frac{4743}{2^{17}}z^{10} + \cdots. \qquad (41.7)$$

If we let $2\sqrt{2}r = R$, reversal of this series gives

$$z = R - \frac{9}{2^6}R^4 + \frac{9}{2^9}R^7 - \frac{9}{2^{12}}R^{10} + \cdots$$

$$= \frac{R(64 - R^3)}{8(8 + R^3)} = \frac{r(2\sqrt{2}-r^3)}{1 + 2\sqrt{2}r^3}. \qquad (41.8)$$

The transformation $\mathrm{sm}\,w = z$ now gives formula (39.3).

Icosahedron. For the icosahedron, we take $\beta = \frac{1}{5}$, $w' = \frac{1}{4}[\sqrt{2}\sqrt{3}\sqrt{(5 + \sqrt{5})} + 3 + \sqrt{5}]$. Then, since

$$\frac{\Gamma(\frac{4}{3})\Gamma(\frac{11}{30})}{\Gamma(\frac{2}{3})\Gamma(\frac{31}{30})} = 2^{5/3}5^{5/6}[\sqrt{2}\sqrt{3}\sqrt{(5 + \sqrt{5})} - 3 - \sqrt{5}], \qquad (41.9)$$

we find

$$2^{1/3}5^{5/6}r = z + \frac{63}{2^45^2}z^4 + \frac{14571}{2^65^5}z^7 + \frac{71\,12133}{2^{11}5^7}z^{10} + \cdots. \qquad (41.10)$$

Putting $2^{1/3}5^{5/6}r = R$, we find the inverse series,

$$z = R - \frac{63}{2^45^2}R^4 + \frac{2637}{2^55^5}R^7 - \frac{5229}{2^45^7}R^{10} + \cdots, \qquad (41.11)$$

whence, from $w = \mathrm{sm}^{-1}z$, we get series (40.6).

V

Conformal Projections Based on Jacobian Elliptic Functions

<u>42. Conformal projection of a circle within a square.</u>

The transformation $\text{sd}\sqrt{2}w = \sqrt{2}s$.   For the case $n = 4$, Schwarz's integral (3.1) can be identified with Jacobian elliptic functions for modulus $1/\sqrt{2}$; that is, it is expressed by

$$\text{sd}(\sqrt{2}w, 1/\sqrt{2}) = \sqrt{2}s \qquad (42.1)$$

by which the interior of the unit circle s is conformally represented by the interior of the square w. The origin is at the centre of the square and the axes are the diagonals. The side of the square is $K = 1 \cdot 854\,075$, and the semidiagonal is $K/\sqrt{2} = 1 \cdot 311\,029$.

Conformal projections of the sphere or parts of the sphere within squares can be easily derived by this transformation. Adams used the mathematically ingenious but laborious method, first described by Guyou, involving elliptic coordinates on the sphere and the use of tables of elliptic integrals. The elliptic coordinates do not form an isometric system, although isometric coordinates can be derived from them. Hence, Adams's reference to elliptic isometric coordinates is misleading.

Computation of coordinates from closed formulae. From (42.1) in the equivalent form,

$$\text{cn}(K - \sqrt{2}w) = x + iy, \qquad (42.2)$$

we get
$$\left.\begin{array}{l} \text{sn}^2(K - \sqrt{2}w) = 1 - (x^2 - y^2) - i2xy, \\[4pt] \text{dn}^2(K - \sqrt{2}w) = \tfrac{1}{2}[1 + (x^2 - y^2) + i2xy]. \end{array}\right\} \qquad (42.3)$$

If we let
$$x^2 + y^2 = A, \qquad x^2 - y^2 = B, \qquad (42.4)$$

then
$$\left.\begin{array}{l} \text{cn}(K - \sqrt{2}w)\,\text{cn}(K - \sqrt{2}\bar{w}) = A, \\[4pt] \text{sn}^2(K - \sqrt{2}w)\,\text{sn}^2(K - \sqrt{2}\bar{w}) = 1 - 2B + A^2, \\[4pt] \text{dn}^2(K - \sqrt{2}w)\,\text{dn}^2(K - \sqrt{2}\bar{w}) = \tfrac{1}{4}(1 + 2B + A^2), \\[4pt] 1 - \tfrac{1}{2}\text{sn}^2(K - \sqrt{2}w)\,\text{sn}^2(K - \sqrt{2}\bar{w}) = \tfrac{1}{2}(1 + 2B - A^2). \end{array}\right\} \qquad (42.5)$$

Then

$$\text{cn}\,2\sqrt{2}u = -\,\text{cn}\,2(K - \sqrt{2}\,u) = -\,\text{cn}\,[(K - \sqrt{2}w) + (K - \sqrt{2}\bar{w})]$$

$$= -\,\frac{2A - \sqrt{(1 + A^2)^2 - 4B^2}}{1 + 2B - A^2}, \qquad (42.6)$$

$$\text{cn}\,2\sqrt{2}v = \frac{1}{\text{cn}\,[(K - \sqrt{2}w) - (K - \sqrt{2}\bar{w})]}$$

$$= \frac{1 + 2B - A^2}{2A + \sqrt{(1 + A^2)^2 - 4B^2}}. \qquad (42.7)$$

Tables for finding u from $\text{cn}\,2\sqrt{2}u$ for modular angle $45°$, based on the *Smithsonian Elliptic Functions Tables*, are given in Tables 10 to 14.

On the circumference of the unit circle, $x^2 + y^2 = 1$, and if we let $x + iy = \cos\theta + i\sin\theta$, the above expressions reduce to

$$-\,\text{cn}\,2\sqrt{2}u = \text{cn}\,2\sqrt{2}v = \tan(45° \pm \theta), \qquad (42.8)$$

taking whichever sign will give a tangent in the range −1 to +1. These results show also that

$$\pm u \pm v = K/\sqrt{2},$$ (42.9)

which are the equations of the four sides of the square.

Computation of coordinates by series. A series for w in powers of z is easily derived from Schwarz's integral (3.1) for $n = 4$,

$$w = \int_0^z (1 - z^4)^{-1/2} \, dz$$

$$= z + \frac{1}{2} \cdot \frac{1}{5} z^5 + \frac{1.3}{2.4} \cdot \frac{1}{9} z^9 + \frac{1.3.5}{2.4.6} \cdot \frac{1}{13} z^{13} + \frac{1.3.5.7}{2.4.6.8} \cdot \frac{1}{17} z^{17} + \cdots,$$ (42.10)

or $w =$

z	$+0 \cdot 007\ 223\ z^{29}$
$+0 \cdot 1 \quad z^5$	$+0 \cdot 005\ 951\ z^{33}$
$+0 \cdot 041\ 667\ z^9$	$+0 \cdot 005\ 013\ z^{37}$
$+0 \cdot 024\ 038\ z^{13}$	$+0 \cdot 004\ 297\ z^{41}$
$+0 \cdot 016\ 085\ z^{17}$	$+0 \cdot 003\ 738\ z^{45}$
$+0 \cdot 011\ 719\ z^{21}$	$+ \cdots,$
$+0 \cdot 009\ 023\ z^{25}$	

(42.11)

which gives 6-figure accuracy for $|z| \not> 0 \cdot 82$. When the computation is done by desk calculator, the closed formulae will be simpler than the use of many terms of this series, and it is recommended only for $|z| \not> 0 \cdot 5$.

By reversing this series, or from $z = \operatorname{sn}\sqrt{2}w / \sqrt{2} \operatorname{dn}\sqrt{2}w$, we get

$$z = w - \frac{1}{10} w^5 + \frac{1}{120} w^9 - \frac{11}{15600} w^{13} + \frac{211}{35\ 36000} w^{17} - \cdots,$$ (42.12)

or $z =$

w	$-0 \cdot 361\ 61\ (w^{29} \times 10^{-7})$
$-1 \cdot 0 \quad (w^5 \times 10^{-1})$	$+0 \cdot 306\ 0 \quad (w^{33} \times 10^{-8})$
$+0 \cdot 833\ 333\ (w^9 \times 10^{-2})$	$-0 \cdot 259\ 0 \quad (w^{37} \times 10^{-9})$
$-0 \cdot 705\ 128\ (w^{13} \times 10^{-3})$	$+0 \cdot 219 \quad (w^{41} \times 10^{-10})$
$+0 \cdot 596\ 719\ (w^{17} \times 10^{-4})$	$-0 \cdot 185 \quad (w^{45} \times 10^{-11})$
$-0 \cdot 504\ 965\ (w^{21} \times 10^{-5})$	$+ \cdots,$
$+0 \cdot 427\ 32 \quad (w^{25} \times 10^{-6})$	

(42.13)

which is adequate for 6-figure accuracy over the whole area of the square.

43. Projection of a hemisphere within a square (Peirce 1877, Guyou 1887, Adams 1925).

Formula in terms of Jacobian functions. The conformal projection of a hemisphere within a square is given by

$$\operatorname{sd}(\sqrt{2}w,\ 1/\sqrt{2}) = \sqrt{2}r,$$ (43.1)

where r is the stereographic projection with $m = \frac{1}{2}$ at the origin. Computation of coordinates is easily done by the closed formulae (42.6) and (42.7) or the series (42.11) with r substituted for z.

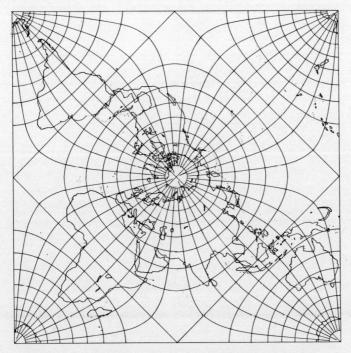

Fig. 28. Projection of a hemisphere within a square,
pole at the centre (sphere within a larger square).

Conformal projections of a hemisphere within a square were
devised by Peirce 1877, Guyou 1887, and Adams 1925, none of whom
used the definition (43.1). It seems not to have been generally
recognized that their projections are merely different aspects of
the same projection, and all of them can be derived in the same
way from different aspects of the stereographic.

Scale. By differentiation of (43.1) we get

$$\frac{dw}{dr} = \frac{dn^2\sqrt{2}w}{cn\sqrt{2}w} = \frac{1}{\sqrt{(1-r^4)}} \tag{43.2}$$

so that the scale of the projection is given by

$$m = \tfrac{1}{2}(1 + p^2 + q^2)/\left|\sqrt{(1-r^4)}\right|. \tag{43.3}$$

At the origin, $m = \tfrac{1}{2}$; at the midpoint of a side, $m = 1/\sqrt{2}$. There
are singular points at the four corners of the square.

*Projection of a hemisphere within a square with the pole at
the centre of the square.* If we start from the stereographic projection with the origin at the pole (16.3), we obtain the projection of Peirce 1877, with the pole at the centre of the square.
Peirce arranged the second hemisphere in four pieces, so that the
whole sphere was mapped within a larger square, as shown in Fig. 28.
This five-piece or quincuncial arrangement led to the description
"quincuncial" being inappropriately applied to the projection it-
self, although other arrangements are possible and have been used.

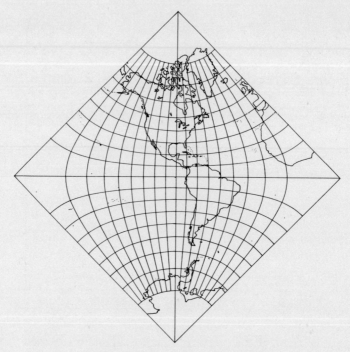

Fig. 29. Projection of a hemisphere within a square,
poles at opposite vertices.

*Projection of a hemisphere within a square with the poles at
opposite vertices.* If we start from the stereographic projection
with the origin on the equator and the poles on the real axis
(15.2), we obtain the projection of Adams 1925, with the poles at
opposite vertices of the square, as shown in Fig. 29.

*Projection of a hemisphere within a square with the poles at
the midpoints of opposite sides.* This is the projection of Guyou
1887, and is also obtained from a stereographic projection with
the origin on the equator but with the poles equidistant from the
axes. It can therefore be computed by letting the stereographic
coordinates (15.2) be turned through $-45°$, that is, they are
multiplied by $(1 - i)/\sqrt{2}$. When the coordinates within a square
have been computed, it is then convenient to turn them through
$+45°$, or to multiply them by $(1 + i)/\sqrt{2}$. The effect of both rota-
tions is merely to change the signs of alternate terms in series
(42.11). That is, using the stereographic coordinates (15.2)
without change, the Guyou projection is given by

$$w = r - 0 \cdot 1 r^5 + 0 \cdot 041\,667\, r^9 - 0 \cdot 024\,038\, r^{13} + \cdots \text{etc.}, \qquad (43.4)$$

where the axes of coordinates are now parallel to the sides of
the square. The inverse series (43.13) similarly undergoes a
change of signs of alternate terms, so that all the coefficients
are positive. The vertices of the square are at latitudes $\pm45°$.

The projection of Guyou, shown in Fig. 30, has often been
described as the doubly-periodic projection, although it is only
one example of a large number of doubly-periodic projections.*

*Doubly-periodic projections discussed in this study are the projections
of the sphere or of a hemisphere within an equilateral triangle, a rhombus,
a square, or a rectangle, and the projection of the sphere upon a regular
tetrahedron.

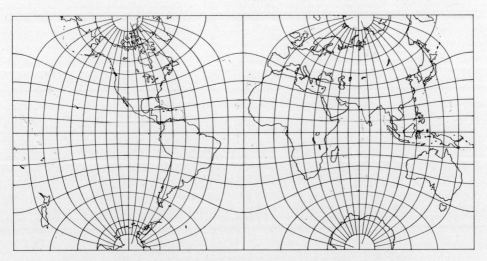

Fig. 30. Projection of a hemisphere within a square, poles at
the midpoints of opposite sides (sphere within a rectangle).

That it is the transverse aspect of the Peirce projection may not
be immediately apparent because Guyou placed two hemispheres side
by side to map the sphere within a rectangle while Peirce divided
the second hemisphere into four pieces so placed as to map the
sphere within a larger square.

    44. Projections of the sphere within a square (Adams 1929,
1936). To map the sphere within a square, we use the Lagrange
projection instead of the stereographic as the parent projection
in equation (42.1). It should be noted, however, that the Lagrange
projection, unlike the stereographic, has two singular points on
the bounding circle, and the character of the projection is changed
as these singular points are placed in different positions on the
boundary of the square. In the first three examples below, the
singular points of the Lagrange projection are mapped at opposite
vertices of the square; the only singular points are those at the
four vertices, and these three projections are different aspects
of the same projection. In the fourth example, the singular points
of the Lagrange projection are mapped at the midpoints of opposite
sides of the square; the resulting projection has six singular
points, and is not the same projection as that in the first three
examples. The singular points of the Lagrange projection could be
placed in other positions on opposite sides of the square, resulting in further projections with different properties.

    *Projection of the sphere within a square, with the poles at
opposite vertices.* In this case, we use the direct Lagrange projection as the parent projection, so that the projection is
defined by

$$\operatorname{sd}(\sqrt{2}\,w,\ 1/\sqrt{2}\,) = \sqrt{2} \tanh \tfrac{1}{4}\zeta. \qquad\qquad (44.1)$$

Coordinates of the Lagrange projection are computed from (17.2),
and computation of the coordinates of the projection within a
square then follows the procedure of Sec. 42. This projection is
illustrated in Fig. 31.

    By differentiation of (44.1) we have

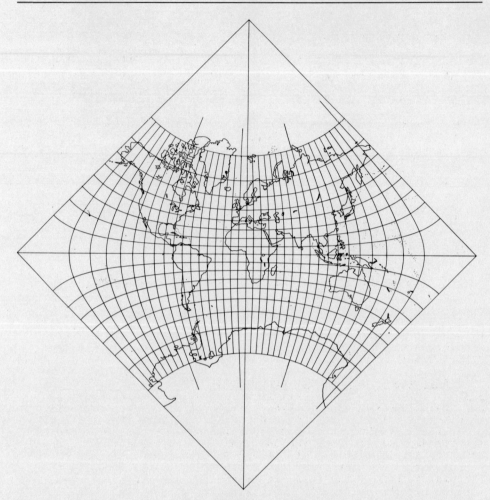

Fig. 31. Projection of the sphere within a square, poles at opposite vertices.

$$\frac{\operatorname{cn}\sqrt{2}w}{\operatorname{dn}^2\sqrt{2}w}\,\mathrm{d}w \;=\; \tfrac{1}{4}\,(1-\tanh^2\tfrac{1}{4}\,\zeta)\,\mathrm{d}\zeta \;=\; \tfrac{1}{4}\,\frac{\operatorname{cn}^2\sqrt{2}w}{\operatorname{dn}^2\sqrt{2}w}\,\mathrm{d}\zeta, \qquad (44.2)$$

whence
$$\mathrm{d}w/\mathrm{d}\zeta \;=\; \tfrac{1}{4}\operatorname{cn}\sqrt{2}w, \qquad\qquad (44.3)$$

from which the scale can be found. The scale at the origin is $m = \tfrac{1}{4}$.

  Projection of the sphere within a square, with the poles symmetrically placed on a diagonal. If the coordinates of the transverse Lagrange projection (20.5) are used in (44.1) instead of those of the direct Lagrange, we get the projection illustrated in Fig. 32, with the poles symmetrically placed on a diagonal of the square. This is the projection of Adams 1929, obtained by him by a much more complicated method of computation.

  Projection of the sphere within a square, with a pole at the centre. By starting with the coordinates of a Lagrange projection

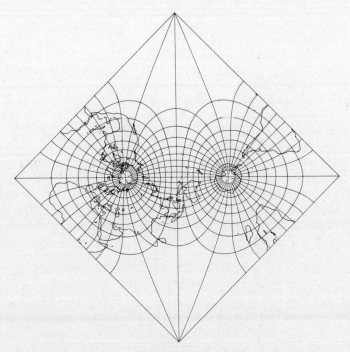

Fig. 32. Projection of the sphere within a square, poles on a diagonal.

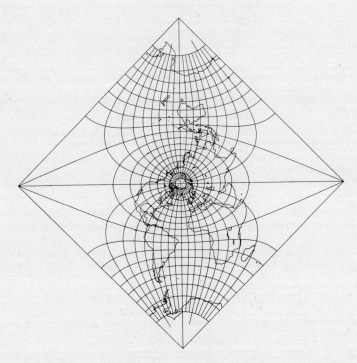

Fig. 33. Projection of the sphere within a square, pole at the centre.

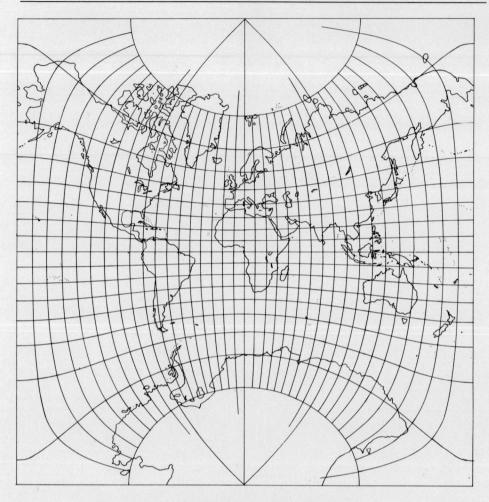

Fig. 34. Projection of the sphere within a square,
poles at the midpoints of opposite sides.

with a pole at the centre (21.4), we can derive a projection of
the sphere within a square, with one pole at the centre of the
square and the other pole represented by two points at opposite
ends of a diagonal. This projection is illustrated in Fig. 33.

*Projection of the sphere within a square, with the poles at
the midpoints of opposite sides.* If the direct Lagrange projection
is rotated through 45°, i.e. if the coordinates (17.2) are multiplied by $(1 - i)/\sqrt{2}$, before the projection within the square is
computed, we get a projection in which the poles are placed at the
midpoints of opposite sides of the square. We can use the same
device as that adopted for the Guyou projection (43.4), changing
the signs of alternate terms in the series (42.11), and getting
coordinates referred to axes parallel to the sides of the square.

This projection is illustrated in Fig. 34. It was derived by
Adams 1936 by a much more complicated method of computation. The
map was used for wall decoration in an airport building at Fort
Worth, Texas.

45. Projection of the sphere upon a cube (Lee 1970).

Derivation of formula. The one-sixth part of the surface of the sphere which is represented by one square face of the cube is a spherical quadrilateral whose diagonals divide it into four spherical triangles, each with angles $\frac{1}{2}\pi$, $\frac{1}{3}\pi$, $\frac{1}{3}\pi$. The stereographic projection of each such spherical triangle is conformally represented upon the infinite half-plane, with origin at the vertex with angle $\frac{1}{2}\pi$, by the function ξ in (23.9), that is,

$$\xi = \frac{12\sqrt{3}r^2(1+r^4)^2}{(1+2\sqrt{3}r^2-r^4)^3}, \qquad 1-\xi = \frac{(1-2\sqrt{3}r^2-r^4)^3}{(1+2\sqrt{3}r^2-r^4)^3}. \qquad (45.1)$$

Representing the infinite half-plane upon the quarter-plane, and then representing the whole plane within the unit circle by (31.1), we can next represent the circle conformally within a square by

$$\mathrm{sd}(\sqrt{2}w,\ 1/\sqrt{2}) = \sqrt{2}[1-\sqrt{(1-\xi)}]/\sqrt{\xi}, \qquad (45.2)$$

the axes of coordinates being the diagonals of the square. By squaring both sides and componendo and dividendo, we get

$$\mathrm{cn}\ \sqrt{2}w = \sqrt{(1-\xi)}, \qquad (45.3)$$

so that one square face of the conformal projection of the sphere upon a cube is given by

$$\mathrm{cn}\sqrt{2}w = \left(\frac{1-2\sqrt{3}r^2-r^4}{1+2\sqrt{3}r^2-r^4}\right)^{3/4}. \qquad (45.4)$$

As $\mathrm{cn}(-u) = \mathrm{cn}\ u$, there is an ambiguity of sign in this definition, but it is easily settled in practice since the origins and quadrants can be made to correspond.

Computation of coordinates from closed formulae. If we let

$$\mathrm{cn}\ \sqrt{2}w = x + iy,$$

then

$$\left.\begin{array}{l} \mathrm{sn}^2\sqrt{2}w = 1-(x^2-y^2)-i2xy, \\ \mathrm{dn}^2\sqrt{2}w = \frac{1}{2}[1+(x^2-y^2)+i2xy]. \end{array}\right\} \qquad (45.5)$$

Also let

$$x^2+y^2 = A, \qquad x^2-y^2 = B. \qquad (45.6)$$

Then

$$\left.\begin{array}{l} \mathrm{cn}\ 2\sqrt{2}u = \mathrm{cn}\ (\sqrt{2}w+\sqrt{2}\bar{w}) = \dfrac{2A-\sqrt{(1+A^2)^2-4B^2}}{1+2B-A^2}, \\[3mm] \mathrm{cn}\ 2\sqrt{2}v = \dfrac{1}{\mathrm{cn}\ (\sqrt{2}w-\sqrt{2}\bar{w})} = \dfrac{1+2B-A^2}{2A+\sqrt{(1+A^2)^2-4B^2}}. \end{array}\right\} \qquad (45.7)$$

On the boundary of the spherical quadrilateral, $x^2 = y^2$, and

$$-\ \mathrm{cn}\ 2\sqrt{2}u = \mathrm{cn}\ 2\sqrt{2}v = (1-2x^2)/(1+2x^2), \qquad (45.8)$$

so that

$$\pm u \pm v = K/\sqrt{2}, \qquad (45.9)$$

which are the equations of the four sides of the square.

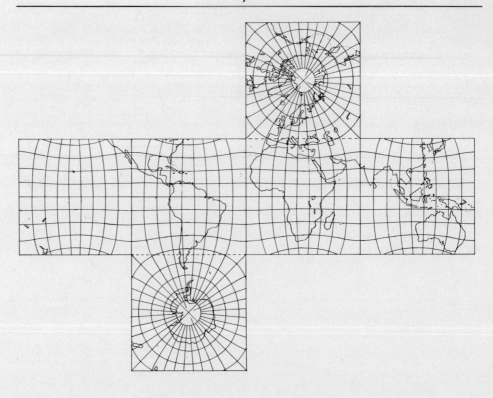

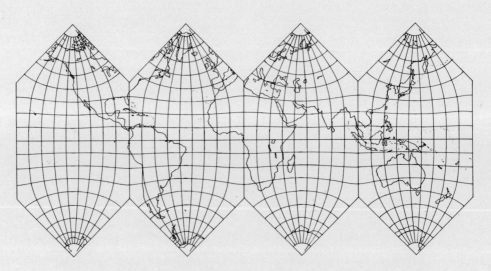

Fig. 35. Projection of the sphere upon a cube, two faces with
a pole at the centre, four faces with centre on the equator.

Computation of coordinates by series. To express the coordinates by a series in powers of r, it is simpler to return to (45.2) and obtain

$$s = \frac{1 - \sqrt{1 - \xi}}{\sqrt{\xi}} = \frac{(1 + 2\sqrt{3}r^2 - r^4)^{3/2} - (1 - 2\sqrt{3}r^2 - r^4)^{3/2}}{2 \cdot 3^{3/4} r (1 + r^4)}$$

$$= 3^{3/4} r (1 - 2r^4 - 10r^{12} - 58r^{16} - 486r^{20} - \cdots). \qquad (45.10)$$

Note that there is a zero term in r^8. From Schwarz's integral (3.1) we have

$$w = \int_0^z (1 - s^4)^{-1/2} ds$$

$$= s + \frac{1}{2} \frac{1}{5} s^5 + \frac{1 \cdot 3}{2 \cdot 4} \frac{1}{9} s^9 + \frac{1 \cdot 3 \cdot 5}{2 \cdot 4 \cdot 6} \frac{1}{13} s^{13} + \cdots. \qquad (45.11)$$

Substituting the value of s and its powers from (45.10) in (45.11), and also substituting

$$R^4 = 3^3 r^4, \qquad \text{or} \qquad R = 2 \cdot 279\,507\,r, \qquad (45.12)$$

we obtain

$$w = R + \frac{7}{270} R^5 + \frac{1}{216} R^9 + \frac{5075}{40\,94064} R^{13} + \cdots, \qquad (45.13)$$

or

$$
\begin{aligned}
w = \quad & R & &+ 0 \cdot 000\,053\,R^{25} \\
& + 0 \cdot 025\,926\,R^5 & &+ 0 \cdot 000\,021\,R^{29} \\
& + 0 \cdot 004\,630\,R^9 & &+ 0 \cdot 000\,009\,R^{33} \\
& + 0 \cdot 001\,240\,R^{13} & &+ 0 \cdot 000\,004\,R^{37} \\
& + 0 \cdot 000\,397\,R^{17} & &+ 0 \cdot 000\,002\,R^{41} \\
& + 0 \cdot 000\,141\,R^{21} & &+ \cdots.
\end{aligned}
\qquad (45.14)
$$

This series is suitable for use within the region where $|R| \not> 1$, which is the greater part of the area of the square face. Both this series and its inverse series are too slowly convergent to be practically useful in the region near a vertex.

Scale. From (45.3) we get

$$\mathrm{sn}^2\sqrt{2}w = 1 - \sqrt{1 - \xi}, \qquad \mathrm{dn}^2\sqrt{2}w = \tfrac{1}{2}(1 + \sqrt{1 - \xi}),$$
$$\mathrm{sn}^2\sqrt{2}w \, \mathrm{dn}^2\sqrt{2}w = \tfrac{1}{2}\xi, \qquad (45.15)$$

and thus we get

$$\mathrm{d\,cn}\sqrt{2}w/dw = -\sqrt{2}\,\mathrm{sn}\sqrt{2}w\,\mathrm{dn}\sqrt{2}w = -\sqrt{\xi}$$
$$= -\frac{2\sqrt{3}\sqrt[4]{3}r(1 + r^4)}{(1 + 2\sqrt{3}r^2 - r^4)^{3/2}}. \qquad (45.16)$$

We also have

$$\frac{d}{dr}\left(\frac{1 - 2\sqrt{3}r^2 - r^4}{1 + 2\sqrt{3}r^2 - r^4}\right)^{3/4} = -\frac{6\sqrt{3}r(1 + r^4)}{(1 + 2\sqrt{3}r^2 - r^4)^{7/4}(1 - 2\sqrt{3}r^2 - r^4)^{1/4}}. \qquad (45.17)$$

From these we get

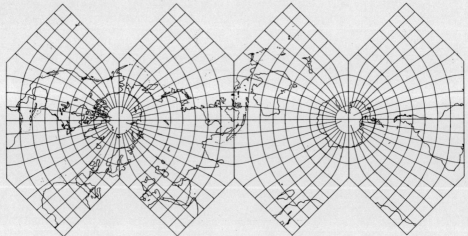

Fig. 36. Projection of the sphere upon a cube, two faces with the equator as a diagonal, four faces with a pole at the midpoint of an edge.

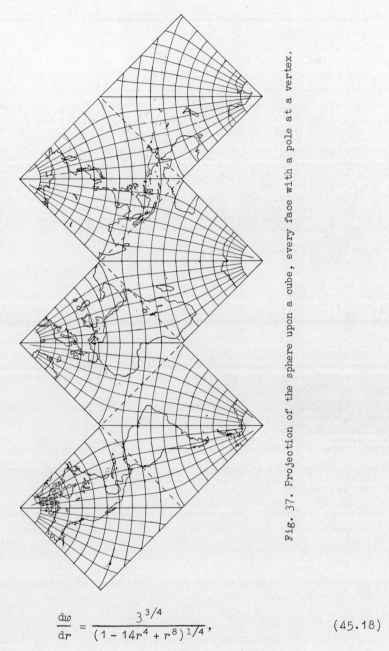

Fig. 37. Projection of the sphere upon a cube, every face with a pole at a vertex.

$$\frac{\mathrm{d}w}{\mathrm{d}r} = \frac{3^{3/4}}{(1 - 14r^4 + r^8)^{1/4}}, \qquad\qquad (45.18)$$

and therefore the scale of the projection is given by

$$m = \tfrac{1}{2}\cdot 3^{1/4}(1 + p^2 + q^2)\cdot\left|(1 - 14r^4 + r^8)^{-1/4}\right|. \qquad (45.19)$$

The scale at the centre of a square face is $\tfrac{1}{2}\cdot 3^{3/4} = 1\cdot139\,754$; at the midpoint of an edge it is $\sqrt{3}/\sqrt{2} = 1\cdot224\,745$. At the midpoint of an edge, the increase in scale over the scale at the origin is therefore $2^{1/2}3^{-1/4}$, or $7\cdot4570\,\%$.

Nature of the projection. Three examples of the projection are shown in Figs. 35 to 37.

In Fig. 35, two faces have a pole at the centre and four faces have a centre on the equator. On the common boundary of two dissimilar faces, $\tan\phi = \cos\lambda$, where λ is reckoned from the central meridian of the equatorial face. The vertices are therefore given by $\tan\phi = \pm1/\sqrt{2}$, or $\phi = \pm35°\ 15'\ 51''\cdot803$. The polar face is derived from the polar stereographic (16.3); the equatorial face is derived from the transverse stereographic (15.2) rotated through 45°, that is, with coordinates multiplied by $(1-i)/\sqrt{2}$.

In Fig. 36, two faces have the equator as a diagonal and four faces have a pole at the midpoint of an edge. On the common boundary between two dissimilar faces, $\tan\phi = \sqrt{2}\sin\lambda - \cos\lambda$, where λ is reckoned from the central meridian of the face containing the pole. The vertices are therefore on the equator at longitudes given by $\tan\lambda = \pm1/\sqrt{2}$, or $\lambda = \pm35°\ 15'\ 51''\cdot803$, and on the meridian $\lambda = \pm\frac{1}{2}\pi$ at latitudes given by $\tan\phi = \pm\sqrt{2}$, or $\phi = \pm54°\ 44'\ 08''\cdot197$.

The graticules of Figs. 35 and 36 are also shown in assemblies in which some faces have been cut along the diagonals.

Fig. 37 shows an example in which all six faces have a similar graticule, with a pole at a vertex. Each face is derived from the oblique stereographic (14.3) with origin given by $\tan\phi_0 = \pm1/\sqrt{2}$, or $\phi_0 = \pm35°\ 15'\ 51''\cdot803$. Two edges represent the meridians $\lambda = \pm60°$, and on the other two edges, $2\sqrt{2}\tan\phi = \sqrt{3}\sin\lambda - \cos\lambda$. If one vertex is the north pole, the opposite vertex is on the central meridian at latitude given by $\sin\phi = -\frac{1}{3}$, or $\phi = -19°\ 28'\ 16''\cdot394$; the other two vertices are at latitudes $+19°\ 28'\ 16''\cdot394$. The equator passes through the midpoints of adjacent edges at longitudes $\pm30°$.

In Fig. 35, the equatorial face is obviously the transverse aspect of the polar face. If we imagine the sphere rolled through a right angle about an axis joining the centres of two opposite equatorial faces, so that a meridian takes up the position formerly occupied by the equator, the effect on the projection is merely to transfer the graticule from one face to another or to rotate the graticule on one face. If the six faces together are considered as one projection of the whole sphere, the graticule of the transverse aspect is identical with that of the direct aspect. The same remarks apply to Fig. 36, where the sphere can be rolled through a right angle so that the pole at the midpoint of an edge is transferred to the midpoint of the opposite edge, and the diagonal representing the equator in one face becomes the other diagonal of the same face.

The graticule of Fig. 37 does not have this property, for the sphere cannot be rolled through a right angle so as to transfer the pole from one vertex to another. It can, however, be rolled through a right angle so as to transfer the pole from a vertex to the midpoint of an edge, and there is a meridian in the graticule of Fig. 36, although not one of the meridians actually drawn, which follows the same course as the equator in Fig. 37.

46. Conformal representations of a circle within a rectangle. The conformal representation of the interior of a circle by the interior of a rectangle can also be effected by means of Jacobian elliptic functions, the dimensions of the rectangle being dependent on the choice of the modulus k. The representation necessarily involves singular points on the boundary, but these may be located in various positions. We shall consider two cases.

Singular points at midpoints of opposite sides. In the representation of the unit circle z given by

$$\sqrt{k'} \, \mathrm{sc} \, \tfrac{1}{2}w = z, \tag{46.1}$$

separation of the real and the imaginary parts gives

$$\left.\begin{aligned}
\frac{\sqrt{k'}\,\mathrm{sn}\,\tfrac{1}{2}u\,\mathrm{cn}\,\tfrac{1}{2}u\,\mathrm{cn'}\,\tfrac{1}{2}v\,\mathrm{dn'}\,\tfrac{1}{2}v}{1 - \mathrm{sn}^2\,\tfrac{1}{2}u\,\mathrm{dn'}^2\,\tfrac{1}{2}v} &= x, \\[2ex]
\frac{\sqrt{k'}\,\mathrm{dn}\,\tfrac{1}{2}u\,\mathrm{sn'}\,\tfrac{1}{2}v}{1 - \mathrm{sn}^2\,\tfrac{1}{2}u\,\mathrm{dn'}^2\,\tfrac{1}{2}v} &= y,
\end{aligned}\right\} \tag{46.2}$$

which show that $x = 0$ when $u = 0$, and $y = 0$ when $v = 0$, so that origins and axes correspond. We also have

$$x^2 + y^2 = \frac{k'(1 - \mathrm{cn}^2\,\tfrac{1}{2}u\,\mathrm{cn'}^2\,\tfrac{1}{2}v)}{1 - \mathrm{sn}^2\,\tfrac{1}{2}u\,\mathrm{dn'}^2\,\tfrac{1}{2}v}. \tag{46.3}$$

On the circumference of the unit circle, $x^2 + y^2 = 1$, and equation (46.3) is then satisfied by $u = \pm K$.

When $v = \pm 2K'$, (46.2) gives

$$x = 0, \qquad y = \pm\sqrt{k'}\,\mathrm{nd}\,\tfrac{1}{2}u. \tag{46.4}$$

At $u = 0$, $v = \pm 2K'$, we therefore have $x = 0$, $y = \pm\sqrt{k'}$.

The transformation (46.1) therefore represents the unit circle within a rectangle of height $2K$ and width $4K'$. The circumference of the circle is represented by $u = \pm K$, the upper and lower sides of the rectangle, and the imaginary axis of the circle is represented by $u = 0$ and also by $v = \pm 2K'$, the left and right sides of the rectangle. At $u = 0$, $v = \pm 2K'$, there are singular points where the imaginary axis of the circle is turned through a right angle, and the upper half of the boundary of the rectangle represents the same line as the lower half.

Singular points at the vertices. Another conformal representation of the unit circle z within a rectangle is given by

$$\sqrt{k'} \, \mathrm{sc}\,\tfrac{1}{2}(K - w) = \frac{1 - z}{1 + z}, \tag{46.5}$$

or, in the inverse form,

$$z = \frac{1 - \sqrt{k'}\,\mathrm{sc}\,\tfrac{1}{2}(K - w)}{1 + \sqrt{k'}\,\mathrm{sc}\,\tfrac{1}{2}(K - w)}. \tag{46.6}$$

If we use the abbreviations, $s = \mathrm{sn}\,\tfrac{1}{2}(K - u)$, $s' = \mathrm{sn'}\,\tfrac{1}{2}v$, etc., separation of the real and the imaginary parts gives

$$\left.\begin{aligned}
x &= \frac{1 - s^2 d'^2 - k'(1 - c^2 c'^2)}{1 - s^2 d'^2 + k'(1 - c^2 c'^2) + 2\sqrt{k'}\,scc'd'}, \\[2ex]
y &= \frac{2\sqrt{k'}\,ds'}{1 - s^2 d'^2 + k'(1 - c^2 c'^2) + 2\sqrt{k'}\,scc'd'}.
\end{aligned}\right\} \tag{46.7}$$

When $u = 0$, then $x = 0$, and when $v = 0$, then $y = 0$, so that origins and axes correspond.

When $u = \pm K$, (46.7) become

$$x = \pm \frac{1 - k's'^2}{1 + k's'^2}, \qquad y = \frac{2\sqrt{k's'}}{1 + k's'^2}, \qquad (46.8)$$

and when $v = \pm 2K'$, they become

$$x = \frac{d^2 - k'}{d^2 + k'}, \qquad y = \pm \frac{2\sqrt{k'}d}{d^2 + k'}, \qquad (46.9)$$

and in both cases, $x^2 + y^2 = 1$. The circumference of the circle is therefore represented by the boundaries of the rectangle, $u = \pm K$, and $v = \pm 2K'$, and there are singular points at the vertices, where

$$x = \pm \frac{1 - k'}{1 + k'}, \qquad y = \pm \frac{2\sqrt{k'}}{1 + k'}. \qquad (46.10)$$

47. Projection of the sphere within a rectangle with a meridian as boundary (Adams 1925).

Definition of the projection. A conformal projection of the sphere within a rectangle, with a meridian as the boundary, and with axes parallel to the sides of the rectangle, is defined by

$$\sqrt{k'} \operatorname{sc} \tfrac{1}{2}(K - w) = \exp\left(-\tfrac{1}{2}\zeta\right), \qquad (47.1)$$

which is the transformation (46.5) with the substitution $s = \tanh \tfrac{1}{4}\zeta$. Adams defined it differently, but his formula may be transformed into this form by changing the origin and axes and by doubling the dimensions.

By separating the real and the imaginary parts, we get

$$\left.\begin{array}{l} \dfrac{\sqrt{k'} \operatorname{sn}\tfrac{1}{2}(K - u)\, \operatorname{cn}\tfrac{1}{2}(K - u)\, \operatorname{cn}'\tfrac{1}{2}v\, \operatorname{dn}'\tfrac{1}{2}v}{1 - \operatorname{sn}^2 \tfrac{1}{2}(K - u)\, \operatorname{dn}'^2\tfrac{1}{2}v} = \exp\left(-\tfrac{1}{2}\psi\right)\cos\tfrac{1}{2}\lambda, \\[3mm] \dfrac{\sqrt{k'} \operatorname{dn}\tfrac{1}{2}(K - u)\, \operatorname{sn}'\tfrac{1}{2}v}{1 - \operatorname{sn}^2 \tfrac{1}{2}(K - u)\, \operatorname{dn}'^2\tfrac{1}{2}v} = \exp\left(-\tfrac{1}{2}\psi\right)\sin\tfrac{1}{2}\lambda, \end{array}\right\} \quad (47.2)$$

and thence, by squaring and adding in the first case and by division in the second, we get

$$\left.\begin{array}{l} \exp\left(-\tfrac{1}{2}\psi\right) = \tan\tfrac{1}{2}c = \dfrac{k'[1 - \operatorname{cn}^2\tfrac{1}{2}(K - u)\, \operatorname{cn}'^2\tfrac{1}{2}v]}{1 - \operatorname{sn}^2\tfrac{1}{2}(K - u)\, \operatorname{dn}'^2\tfrac{1}{2}v}, \\[3mm] \tan\tfrac{1}{2}\lambda = \dfrac{\operatorname{dn}\tfrac{1}{2}(K - u)\, \operatorname{sn}'\tfrac{1}{2}v}{\operatorname{sn}\tfrac{1}{2}(K - u)\, \operatorname{cn}\tfrac{1}{2}(K - u)\, \operatorname{cn}'\tfrac{1}{2}v\, \operatorname{dn}'\tfrac{1}{2}v}. \end{array}\right\} \quad (47.3)$$

These formulae are amenable to some further simplification, using some of the many formulae relating to Jacobian functions. Thus

$$\tan\tfrac{1}{2}\phi = \frac{1 - \tan\tfrac{1}{2}c}{1 + \tan\tfrac{1}{2}c}$$

$$= \frac{1 - \operatorname{sn}^2\tfrac{1}{2}(K - u)\, \operatorname{dn}'^2\tfrac{1}{2}v - k'[1 - \operatorname{cn}^2\tfrac{1}{2}(K - u)\, \operatorname{cn}'^2\tfrac{1}{2}v]}{1 - \operatorname{sn}^2\tfrac{1}{2}(K - u)\, \operatorname{dn}'^2\tfrac{1}{2}v + k'[1 - \operatorname{cn}^2\tfrac{1}{2}(K - u)\, \operatorname{cn}'^2\tfrac{1}{2}v]}$$

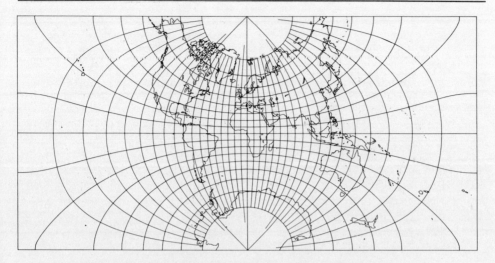

Fig. 38. Projection of the sphere within a rectangle
bounded by a meridian, for $k = \sin 45°$.

$$= \frac{1 + k'}{1 - k'} \cdot \frac{dn^2 \frac{1}{2}(K - u) - k'}{dn^2 \frac{1}{2}(K - u) + k'} \cdot \frac{1 - k'sn'^2 \frac{1}{2}v}{1 + k'sn'^2 \frac{1}{2}v}$$

$$= \sqrt{\left[\frac{1 - dn\, u}{1 + dn\, u} \cdot \frac{1 + k'sn'(K' - v)}{1 - k'sn\,(K' - v)}\right]} \qquad (47.4)$$

$$\tan \tfrac{1}{2}\lambda = \sqrt{\left[\frac{1 + dn\,(K - u)}{1 - dn\,(K - u)} \cdot \frac{1 - sn'(K' - v)}{1 + sn'(K' - v)}\right]}$$

$$= \sqrt{\left[\frac{dn\, u + k'}{dn\, u - k'} \cdot \frac{1 - sn'(K' - v)}{1 + sn'(K' - v)}\right]} \qquad (47.5)$$

and thence

$$\left. \begin{array}{l} \cos \phi = \dfrac{1 - \tan^2 \frac{1}{2}\phi}{1 + \tan^2 \frac{1}{2}\phi} = \dfrac{dn\, u - k'sn'(K' - v)}{1 - k'dn\, u\, sn'(K' - v)}, \\[2ex] \cos \lambda = \dfrac{1 - \tan^2 \frac{1}{2}\lambda}{1 + \tan^2 \frac{1}{2}\lambda} = \dfrac{dn\, u\, sn'(K' - v) - k'}{dn\, u - k'sn'(K' - v)}. \end{array} \right\} \qquad (47.6)$$

Nature of the projection. The expressions (47.6) are the
formulae for inverse computation of latitude and longitude from
projection coordinates, and enable the nature of the projection
to be studied.

When $u = 0$, then $\phi = 0$, so that the v-axis represents the
equator. When $v = 0$, then $\lambda = 0$, so that the u-axis represents
the zero meridian. When $u = \pm K$ or when $v = \pm 2K'$, then $\lambda = \pm \pi$.
The sphere is therefore mapped within a rectangle of height $2K$
and width $4K'$, the boundary of the rectangle being the meridian
$\lambda = \pm \pi$. The poles are located at the midpoints of the upper and
lower sides of the rectangle. At the vertices of the rectangle,
$\tan \frac{1}{2}c = k'$.

By differentiation of (47.1) we get

$$dw/d\zeta = \operatorname{sn} \tfrac{1}{2}(K-w) \operatorname{sn} \tfrac{1}{2}(K+w) = \operatorname{cn} w /(\operatorname{dn} w + k'). \tag{47.7}$$

From this we find that the scale at the origin is $1/(1+k')$. We also find that $|dw/d\zeta|$ is zero at the poles and infinite at the vertices of the rectangle, so that these are singular points of the projection.

For a rectangle in which the length of the equator is twice the length of the central meridian, we take $K = K'$, or $k = k' = \sin 45°$. This projection is illustrated in Fig. 38.

Equation (47.1) can be written in the equivalent form,

$$\frac{1 + \operatorname{sc} \tfrac{1}{2}w \operatorname{dn} \tfrac{1}{2}w}{\operatorname{dn} \tfrac{1}{2}w - k' \operatorname{sc} \tfrac{1}{2}w} = \exp \tfrac{1}{2}\zeta. \tag{47.8}$$

For $k = 1$, $k' = 0$, the elliptic functions become hyperbolic functions, and this expression becomes

$$\cosh \tfrac{1}{2}w + \sinh \tfrac{1}{2}w = \exp \tfrac{1}{2}\zeta, \tag{47.9}$$

or $w = \zeta$, which defines the Mercator projection. We now have $K' = \tfrac{1}{2}\pi$ and $K = \infty$, so that the rectangle is of width 2π and of infinite height.

Computation of coordinates. Adams described a method of computing the coordinates of points in the interior of his projection within a rectangle by introducing a number of auxiliary quantities. A much simpler approach is possible by the use of equations (47.6) which, with obvious abbreviations, can be written

$$\cos \phi = \frac{d - k's'}{1 - k'ds'}, \qquad \cos \lambda = \frac{ds' - k'}{d - k's'}, \tag{47.10}$$

whence

$$\cos \phi \cos \lambda = \frac{ds' - k'}{1 - k'ds'}. \tag{47.11}$$

From (47.11) we get

$$ds' = \frac{k' + \cos \phi \cos \lambda}{1 + k' \cos \phi \cos \lambda} = A, \tag{47.12}$$

and from (47.12) in conjunction with either of equations (47.10),

$$d - k's' = \frac{k^2 \cos \phi}{1 + k' \cos \phi \cos \lambda} = B. \tag{47.13}$$

Solution of equations (47.12) and (47.13) finally gives

$$\left. \begin{aligned} \operatorname{dn} u &= \tfrac{1}{2}[\sqrt{(B^2 + 4k'A)} + B], \\ k' \operatorname{sn}'(K' - v) &= \tfrac{1}{2}[\sqrt{(B^2 + 4k'A)} - B]. \end{aligned} \right\} \tag{47.14}$$

On the equator, for $u = 0$, $\psi = 0$, the second of equations (47.2) gives

$$\operatorname{sn}' \tfrac{1}{2}v = \frac{1 + k' - \sqrt{(1 + k'^2 - 2k' \cos \lambda)}}{2k' \sin \tfrac{1}{2}\lambda}. \tag{47.15}$$

On the central meridian,

$$\operatorname{sc}\tfrac{1}{2}(K-u) = \sqrt{}\tan\tfrac{1}{2}c/\sqrt{k'}, \tag{47.16}$$

and on the boundaries,

$$u = K, \qquad \operatorname{sn}'\tfrac{1}{2}v = \sqrt{}\tan\tfrac{1}{2}c/\sqrt{k'}, \tag{47.17}$$

$$v = 2K', \qquad \operatorname{dn}\tfrac{1}{2}(K-u) = \sqrt{k'}/\sqrt{}\tan\tfrac{1}{2}c. \tag{47.18}$$

48. Projection of the sphere within a rectangle with singular points on the equator (Lee 1965).

Definition of the projection. A conformal projection of the sphere within a rectangle, without singular points at the vertices, is defined by

$$\sqrt{k'}\,\operatorname{sc}\tfrac{1}{2}w = \tanh\tfrac{1}{4}\zeta. \tag{48.1}$$

which is transformation (46.1), s denoting the Lagrange projection. Separating the real and the imaginary parts, we get

$$\left. \begin{aligned}
\frac{\sqrt{k'}\operatorname{sn}\tfrac{1}{2}u\,\operatorname{cn}\tfrac{1}{2}u\,\operatorname{cn}'\tfrac{1}{2}v\,\operatorname{dn}'\tfrac{1}{2}v}{1 - \operatorname{sn}^2\tfrac{1}{2}u\,\operatorname{dn}'^2\tfrac{1}{2}v} &= \frac{\sinh\tfrac{1}{2}\psi}{\cosh\tfrac{1}{2}\psi + \cos\tfrac{1}{2}\lambda} = g, \\[2ex]
\frac{\sqrt{k'}\operatorname{dn}\tfrac{1}{2}u\,\operatorname{sn}'\tfrac{1}{2}v}{1 - \operatorname{sn}^2\tfrac{1}{2}u\,\operatorname{dn}'^2\tfrac{1}{2}v} &= \frac{\sin\tfrac{1}{2}\lambda}{\cosh\tfrac{1}{2}\psi + \cos\tfrac{1}{2}\lambda} = h,
\end{aligned} \right\} \tag{48.2}$$

whence we get

$$g^2 + h^2 = \frac{k'(1 - \operatorname{cn}^2\tfrac{1}{2}u\,\operatorname{cn}'^2\tfrac{1}{2}v)}{1 - \operatorname{sn}^2\tfrac{1}{2}u\,\operatorname{dn}'^2\tfrac{1}{2}v} = \frac{\cosh\tfrac{1}{2}\psi - \cos\tfrac{1}{2}\lambda}{\cosh\tfrac{1}{2}\psi + \cos\tfrac{1}{2}\lambda}. \tag{48.3}$$

From this we also get

$$\tan\tfrac{1}{2}\phi = \tanh\tfrac{1}{2}\psi = \frac{2g}{1 + g^2 + h^2}$$

$$= \frac{2\sqrt{k'}\operatorname{sn}\tfrac{1}{2}u\,\operatorname{cn}\tfrac{1}{2}u\,\operatorname{cn}'\tfrac{1}{2}v\,\operatorname{dn}'\tfrac{1}{2}v}{1 - \operatorname{sn}^2\tfrac{1}{2}u\,\operatorname{dn}'^2\tfrac{1}{2}v + k'(1 - \operatorname{cn}^2\tfrac{1}{2}u\,\operatorname{cn}'^2\tfrac{1}{2}v)}, \tag{48.4}$$

$$\tan\tfrac{1}{2}\lambda = \frac{2h}{1 - g^2 - h^2}$$

$$= \frac{2\sqrt{k'}\operatorname{dn}\tfrac{1}{2}u\,\operatorname{sn}'\tfrac{1}{2}v}{1 - \operatorname{sn}^2\tfrac{1}{2}u\,\operatorname{dn}'^2\tfrac{1}{2}v - k'(1 - \operatorname{cn}^2\tfrac{1}{2}u\,\operatorname{cn}'^2\tfrac{1}{2}v)}. \tag{48.5}$$

After some further manipulation, these become

$$\tan\tfrac{1}{2}\phi = 2\sqrt{k'}(1 + k') \cdot \frac{\operatorname{sn}\tfrac{1}{2}u\,\operatorname{cn}\tfrac{1}{2}u}{\operatorname{dn}^2\tfrac{1}{2}u + k'} \cdot \frac{\operatorname{cn}'\tfrac{1}{2}v\,\operatorname{dn}'\tfrac{1}{2}v}{1 + k'\operatorname{sn}'^2\tfrac{1}{2}v}$$

$$= \sqrt{\left[k'\,\frac{1 - \operatorname{dn}u}{\operatorname{dn}u + k'} \cdot \frac{1 + \operatorname{sn}'(K' - v)}{1 - k'\operatorname{sn}'(K' - v)}\right]} \tag{48.6}$$

86

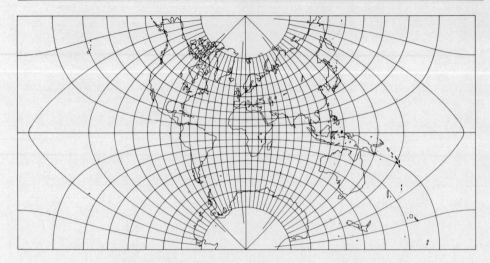

Fig. 39. Projection of the sphere within a rectangle, with singular points at the midpoints of the sides, for $k = \sin 45°$.

$$\tan \tfrac{1}{2}\lambda = 2\sqrt{k'(1+k')} \cdot \frac{\operatorname{dn}\tfrac{1}{2}u}{\operatorname{dn}^2\tfrac{1}{2}u - k'} \cdot \frac{\operatorname{sn}'\tfrac{1}{2}v}{1 - k'\operatorname{sn}'^2\tfrac{1}{2}v}$$

$$= \sqrt{\left[k' \frac{1+\operatorname{dn}u}{\operatorname{dn}u - k'} \cdot \frac{1 - \operatorname{sn}'(K'-v)}{1 + k'\operatorname{sn}'(K'-v)} \right]}. \qquad (48.7)$$

Then, from

$$\cos\phi = \frac{1 - \tan^2\tfrac{1}{2}\phi}{1 + \tan^2\tfrac{1}{2}\phi}, \qquad \cos\lambda = \frac{1 - \tan^2\tfrac{1}{2}\lambda}{1 + \tan^2\tfrac{1}{2}\lambda}, \qquad (48.8)$$

we derive

$$\left. \begin{aligned} \cos\phi &= \frac{(1+k')[\operatorname{dn}u - k'\operatorname{sn}'(K'-v)]}{(1-k')[\operatorname{dn}u + k'\operatorname{sn}'(K'-v)] + 2k'[1 - \operatorname{dn}u\operatorname{sn}'(K'-v)]}, \\ \cos\lambda &= \frac{(1-k')[\operatorname{dn}u + k'\operatorname{sn}'(K'-v)] - 2k'[1 - \operatorname{dn}u\operatorname{sn}'(K'-v)]}{(1+k')[\operatorname{dn}u - k'\operatorname{sn}'(K'-v)]} \end{aligned} \right\} (48.9)$$

which are the formulae for inverse computation of latitude and longitude from projection coordinates.

Nature of the projection. From the original definition (48.1) we see that when $u = 0$, then $\phi = 0$, and when $v = 0$, then $\lambda = 0$, so that the v-axis represents the equator and the u-axis represents the zero meridian.

From (48.9), when $u = \pm K$, then $\lambda = \pm\pi$, and when $v = \pm 2K'$, then $\phi = 0$. The equator is therefore represented by the three straight lines, $u = 0$ and $v = \pm 2K'$, and at their intersections we have $\tan\tfrac{1}{4}\lambda = \pm\sqrt{k'}$.

The sphere is mapped within a rectangle of height $2K$ and length $4K'$, with the poles located at the midpoints of the upper and lower sides, which represent the meridian $\lambda = \pm\pi$. The equator is represented by the v-axis and by the left and right sides of the rectangle.

By differentiation of (48.1) we get

$$\frac{dw}{d\zeta} = \frac{dn\,\frac{1}{2}w - dn\,(K - \frac{1}{2}w)}{2\sqrt{k'(1 - k')}}. \tag{48.10}$$

From this we find that $|dw/d\zeta|$ is zero at $u = \pm K$, $v = 0$, and infinite at $u = 0$, $v = \pm 2K'$, so that the singular points are the midpoints of the four sides of the rectangle. The scale at the origin is $1/(2\sqrt{k'})$.

This projection is illustrated in Fig. 39 for the case $k = k' = \sin 45°$.

Computation of coordinates. Equations (48.9) can be written, with obvious abbreviations, as

$$\left.\begin{aligned}
\cos\phi &= \frac{(1 + k')(d - k's')}{(1 - k')(d + k's') + 2k'(1 - ds')}, \\[2mm]
\cos\lambda &= \frac{(1 - k')(d + k's') - 2k'(1 - ds')}{(1 + k')(d - k's')},
\end{aligned}\right\} \tag{48.11}$$

whence

$$\cos\phi\,\cos\lambda = \frac{(1 - k')(d + k's') - 2k'(1 - ds')}{(1 - k')(d + k's') + 2k'(1 - ds')}, \tag{48.12}$$

and thence

$$\left.\begin{aligned}
\frac{1 + \cos\phi\,\cos\lambda}{1 - \cos\phi\,\cos\lambda} &= \frac{(1 - k')(d + k's')}{2k'(1 - ds')}, \\[2mm]
\frac{\cos\phi}{1 - \cos\phi\,\cos\lambda} &= \frac{(1 + k')(d - k's')}{4k'(1 - ds')}.
\end{aligned}\right\} \tag{48.13}$$

From these equations, we get

$$\frac{d - k's'}{d + k's'} = \frac{2(1 - k')\cos\phi}{(1 + k')(1 + \cos\phi\,\cos\lambda)}, \tag{48.14}$$

whence

$$\frac{d}{k's'} = \frac{(1 + k')(1 + \cos\phi\,\cos\lambda) + 2(1 - k')\cos\phi}{(1 + k')(1 + \cos\phi\,\cos\lambda) - 2(1 - k')\cos\phi}. \tag{48.15}$$

Substitution of this expression in either of equations (48.13) now produces quadratic equations in d and $k's'$. Thus, if we put

$$\left.\begin{aligned}
A &= k^2(1 - \cos\phi\,\cos\lambda), \\
B &= 2(1 - k')\cos\phi - (1 + k')(1 + \cos\phi\,\cos\lambda), \\
C &= 2(1 - k')\cos\phi + (1 + k')(1 + \cos\phi\,\cos\lambda),
\end{aligned}\right\} \tag{48.16}$$

the equations are

$$\left.\begin{aligned}
Bd^2 - Ad + k'C &= 0, \\
Ck'^2s'^2 + Ak's' + k'B &= 0,
\end{aligned}\right\} \tag{48.17}$$

to which the solutions are

$$\begin{aligned}
\mathrm{dn}\, u &= [A - \sqrt{(A^2 - 4k'BC)}]/2B, \\
k'\mathrm{sn}'(K' - v) &= -[A - \sqrt{(A^2 - 4k'BC)}]/2C.
\end{aligned} \right\} \quad (48.18)$$

Simplified formulae for the coordinates on the axes and on the boundaries are

$$\begin{aligned}
u &= 0, & \mathrm{sn}'\tfrac{1}{2}v &= \tan \tfrac{1}{4}\lambda/\sqrt{\sqrt{k'}}, \\
v &= 0, & \mathrm{sc}\tfrac{1}{2}u &= \tanh \tfrac{1}{4}\psi/\sqrt{k'}, \\
v &= 2K', & \mathrm{dn}\tfrac{1}{2}u &= \sqrt{k'}/\tan \tfrac{1}{4}\lambda.
\end{aligned} \right\} \quad (48.19)$$

When $u = K$ and $\lambda = \pi$, the second of equations (48.2) gives

$$\mathrm{sn}'\tfrac{1}{2}v = \frac{(1 + k')\cosh \tfrac{1}{2}\psi - \sqrt{[(1 + k')^2 \cosh^2 \tfrac{1}{2}\psi - 4k']}}{2k'}. \quad (48.20)$$

49. Projections of a hemisphere within a rectangle. For a projection of the same dimensions as those given above for the sphere within a rectangle, but covering only a hemisphere, we write 2ζ instead of ζ. The formulae already given will apply to the new projections with the substitution of λ for $\tfrac{1}{2}\lambda$ and of $\cos^2 \phi/(1 + \sin^2 \phi)$ for $\cos \phi$.

These projections are not considered of sufficient interest to be illustrated, but two examples are briefly treated below.

Projection of a hemisphere within a rectangle bounded by a meridian (Lee 1965). The projection of the sphere within a rectangle defined by (47.1) becomes

$$\sqrt{k'}\,\mathrm{sc}\tfrac{1}{2}(K - w) = \exp(-\zeta) \quad (49.1)$$

when applied to a hemisphere. The v-axis represents the equator, the u-axis represents the zero meridian, and the boundaries of the rectangle, $u = K$ and $v = 2K'$, represent the meridian $\lambda = \tfrac{1}{2}\pi$.

For case $K = 2K'$, or $k' = (\sqrt{2} - 1)/(\sqrt{2} + 1)$, the hemisphere is mapped within a square, and the projection is identical with that of Guyou, described in Sec. 43 above and shown in Fig. 29.

For $k = 0$, $k' = 1$, equation (49.1) becomes

$$\tan \tfrac{1}{2}(\tfrac{1}{2}\pi - w) = \exp(-\zeta), \quad \text{or} \quad w = \mathrm{lam}^{-1}\zeta, \quad (49.2)$$

which defines the transverse Mercator projection.

Projection of a hemisphere within a rectangle bounded by the equator (Lee 1965). The projection of the sphere within a rectangle defined by (48.1) becomes

$$\sqrt{k'}\,\mathrm{sc}\tfrac{1}{2}w = \tanh \tfrac{1}{2}\zeta \quad (49.3)$$

when applied to a hemisphere. The equator is represented by the lines, $u = 0$ and $v = \pm 2K'$, and the meridian $\lambda = \tfrac{1}{2}\pi$ is represented by the line $u = K$. The equator is also represented by the line $u = 2K$, so that if we extend the projection to this line, that is, if we add another rectangle, the hemisphere is represented by a rectangle bounded by the equator, with the pole at the centre.

If we take $K = 2K'$, the rectangle becomes a square, and the projection is that of Peirce (Fig. 28), referred to a new origin and new axes.

For $k = 0$, $k' = 1$, equation (49.3) becomes $\tan \tfrac{1}{2}w = \tanh \tfrac{1}{2}\zeta$, which defines the transverse Mercator projection.

50. Conformal projection of a rectangle within an ellipse.
A conformal representation of an ellipse within a circle was described by Schwarz 1869c, the formula implying a rectangle as an
intermediate transformation. That is, a rectangle w, of height $2K$
and width $4K'$, can be conformally represented within an ellipse z
by

$$z = \sinh(\pi w/4K').\tag{50.1}$$

By separating the real and the imaginary parts, we get

$$x = \sinh\frac{\pi u}{4K'}\cos\frac{\pi v}{4K'}, \qquad y = \cosh\frac{\pi u}{4K'}\sin\frac{\pi v}{4K'},\tag{50.2}$$

from which

$$\frac{x^2}{\sinh^2(\pi u/4K')} + \frac{y^2}{\cosh^2(\pi u/4K')} = 1,\tag{50.3}$$

so that any ordinate of abscissa u is represented by an ellipse of
which the semiaxes are $\sinh(\pi u/4K')$ and $\cosh(\pi u/4K')$, and the
foci are $x = 0$, $y = \pm 1$. Thus, all the ordinates are represented
by confocal ellipses.

The boundaries $u = \pm K$ are represented by an ellipse of semiaxes $\sinh(\pi K/4K')$ and $\cosh(\pi K/4K')$. The vertices of the rectangle, $u = \pm K$, $v = \pm 2K'$, are represented by the extremities of
the major axis of the ellipse. The extremities of the v-axis of
the rectangle are represented by the foci. The left and right
sides of the rectangle are represented by those portions of the
major axis between the foci and the extremities; these are cuts in
the representation, and the areas above and below them are not
connected.

For the major axis of the ellipse to be twice the minor axis,
we must have $\sinh(\pi K/4K') = 1/\sqrt{3}$, whence $K/K' = 0.699\,398$, and we
then find $k = \sin 23°·8958$. To agree with the projection of Adams,
and to make use of available tables of elliptic functions, we have
instead taken $k = \sin 25°$, for which the semiaxes of the ellipse
are $0·5908$ and $1·1615$.

51. Projection of the sphere within a cut ellipse (Adams 1925).
Adams described a conformal projection of the sphere within an
ellipse, which is defined by

$$z = \sinh\left[\frac{\pi}{4K'}\left(K - 2\,\mathrm{sc}^{-1}\frac{\exp(-\tfrac{1}{2}\zeta)}{\sqrt{k'}}\right)\right],\tag{51.1}$$

which is equivalent to the two transformations,

$$\sqrt{k'}\,\mathrm{sc}\tfrac{1}{2}(K - w) = \exp(-\tfrac{1}{2}\zeta), \qquad z = \sinh(\pi w/4K').\tag{51.2}$$

The first of these is the projection of the sphere within a rectangle with singular points on the bounding meridian (47.1), the
second is a representation of this rectangle within an ellipse.

By differentiation of the second of equations (51.2) we get

$$\frac{dz}{dw} = \frac{\pi}{4K'}\cosh\frac{\pi w}{4K'},\tag{51.3}$$

so that, with (47.7), we get

$$\frac{dz}{d\zeta} = \frac{\pi}{4K'}\cdot\frac{\mathrm{cn}\,w}{\mathrm{dn}\,w + k'}\cosh\frac{\pi w}{4K'}.\tag{51.4}$$

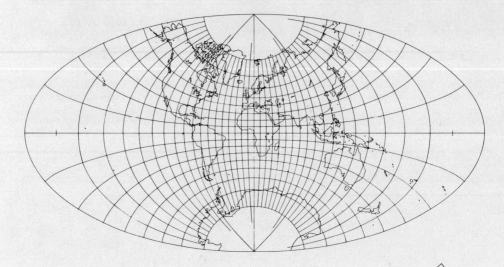

Fig. 40. Projection of the sphere within an
ellipse, with singular points on the bounding
meridian.

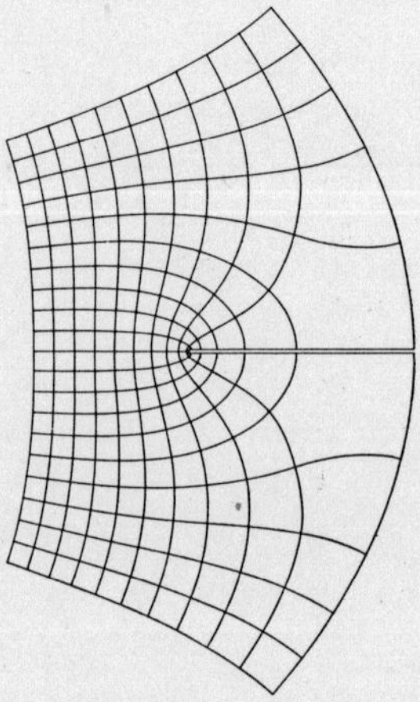

Fig. 41. Graticule at 1° interval in the region
near the extremity of the major axis in the
projection of Fig. 40.

From this we find that $|dz/d\zeta|$ is zero at $u = \pm K$, $v = 0$, infinite
at $u = \pm K$, $v = \pm 2K'$, and zero at $u = 0$, $v = \pm 2K'$, so that there
are singular points of the projection at the poles, the extremities of the major axis, and the foci. There are slits along the
major axis from each extremity to the nearer focus.

Adams mentioned only the singular points at the poles and was
apparently unaware of the others. His table of coordinates wrongly
gives the intersection of the equator and the bounding meridian as
the extremity of the major axis instead of the focus.

The projection, for $k = \sin 25°$, is illustrated in Fig. 40,
and the graticule in the region near the extremity of the major
axis is shown in Fig. 41, where, of course, the width of the slit
is exaggerated.

52. Projection of the sphere within an ellipse without cuts
(Gougenheim 1950, Lee 1965). Another conformal projection of the
sphere within an ellipse is defined by

$$z = \sinh\left[\frac{\pi}{2K'}\, sc^{-1}\, \frac{\tanh\frac{1}{4}\zeta}{\sqrt{k'}}\right],\qquad (52.1)$$

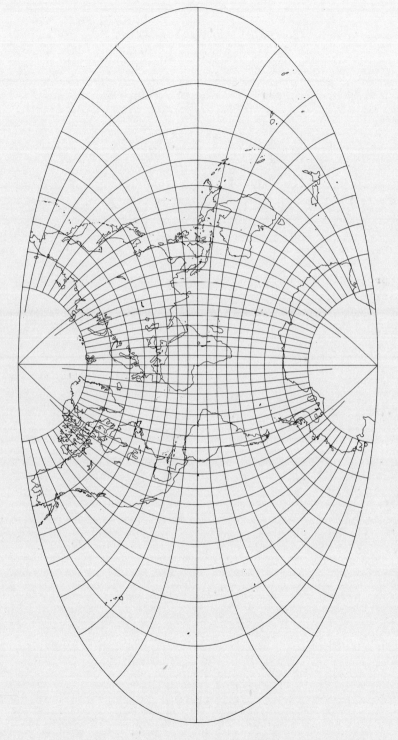

Fig. 42. Projection of the sphere within an ellipse, with singular points only at the poles.

which is equivalent to the two transformations,

$$\sqrt{k'} \operatorname{sc} \tfrac{1}{2}w = \tanh \tfrac{1}{4}\zeta, \qquad s = \sinh (\pi w/4K'). \tag{52.2}$$

The first of these is the projection of the sphere within a rectangle with singular points on the equator (48.1), and the second
is a representation of this rectangle within an ellipse. The
singular points on the equator are cancelled in the second transformation, and the only singular points remaining are those at
the poles.

From (51.3) and (48.10) we now have

$$\frac{ds}{d\zeta} = \frac{\pi[\operatorname{dn} \tfrac{1}{2}w - \operatorname{dn}(K - \tfrac{1}{2}w)]}{8K'\sqrt{k'}(1 - k')} \cosh \frac{\pi w}{4K'}, \tag{52.3}$$

from which we find that $|ds/d\zeta|$ is zero at $u = \pm K$, $v = 0$, so that
the poles are singular points of the projection. There are no
singular points at the extremities of the major axis or at the foci.
In the latter case, $|ds/d\zeta|$ takes the form $0 \times \infty$, but it can be
shown (by expansion in series, for example) that it is finite and
non-zero. The scale at the origin is $\pi/8K'\sqrt{k'}$.

The projection, for $k = \sin 25°$, is illustrated in Fig. 42.

Equation (52.1) was first used as the definition of a projection by Gougenheim 1950, who however believed that it was
merely another formulation of Adams's projection (51.1). Lee 1965,
in ignorance of Gougenheim's work, noticed the additional singular
points which Adams had overlooked, and derived equation (52.1) to
define a projection without these singular points.

53. Transverse Mercator projections of the entire spheroid.
The definition (12.3) can be applied to a projection of the spheroid, using the spheroidal isometric latitude (9.4) instead of the
spherical isometric latitude (4.4), and the resulting projection
has every right to be called a transverse Mercator projection of
the spheroid. It is not, however, the only projection of the
spheroid entitled to this name. The transverse Mercator projection
of the sphere possesses a number of properties, not all of which
can be possessed by any one projection of the spheroid. Instead,
there are a number of projections (Gdowski 1964 asserted that there
is an infinity of such projections), each corresponding in one way
or another to the transverse Mercator projection of the sphere,
and all becoming identical with that projection when the eccentricity of the spheroid is put equal to zero. Of three such projections so far investigated, one, known as the Gauss-Krüger projection, preserves a constant scale along the central meridian; in
the spheroidal projection given by (12.3), the scale along the
central meridian increases from the equator towards the pole; in
the Thompson projection the scale along the central meridian
decreases from the equator towards the pole. The projection of
(12.3), like the transverse Mercator projection of the sphere,
represents the point, $\psi = 0$, $\lambda = \tfrac{1}{2}\pi$, at infinity; the other two
projections represent the entire spheroid within a finite area.

The coordinates of the Gauss-Krüger form of the transverse
Mercator projection of the spheroid have been expressed as infinite
series in powers of the longitude reckoned from the central meridian (or in powers of the complex variable ζ), and these are adequate for the widths of the belts over which the projection has
been used in geodetic survey. If the projection is to include
the entire spheroid, however, it is necessary to seek formulae
in closed form.

The coordinates are expressed in terms of the isometric latitude and longitude by the equation, $x + iy = f(\psi + i\lambda)$. When $\lambda = 0$, then $y = 0$ and $x = f(\psi)$. If the scale is to be correct along the length of the initial meridian, we must therefore have $f(\psi) = M$, where M is the length of the arc of the meridian from the origin (on the equator) to the point being projected. But M cannot be expressed in terms of ψ in closed form by means of any known functions. It is possible, however, to express both (x, y) and (ψ, λ) in closed form in terms of a third set of isometric parameters (u, v) by means of Jacobian elliptic functions. This method of attack is due to Professor E. H. Thompson, who established the transformation formulae in 1945 but did not publish them and did not pursue the matter further. Subsequent investigation of the graticule was by Lee 1962.

54. The Thompson transverse Mercator projection of the entire spheroid (Thompson 1945). Let the u-axis correspond to the initial meridian or ψ-axis, and on this axis put

$$\sin\phi = \text{sn}\,(u, k), \tag{54.1}$$

where the modulus k is equal to the eccentricity of the spheroid. It follows that

$$\cos\phi = \text{cn}\,u, \qquad 1 - k^2\sin^2\phi = \text{dn}^2 u, \qquad d\phi = \text{dn}\,u\,du, \tag{54.2}$$

and thence, from (9.1),

$$\rho = ak'^2\text{nd}^3 u, \qquad \nu = a\,\text{nd}\,u. \tag{54.3}$$

The isometric latitude (9.4) can now be expressed as

$$\psi = \tanh^{-1}(\text{sn}\,u) - k\tanh^{-1}(k\,\text{sn}\,u). \tag{54.4}$$

To express (ψ, λ) in terms of (u, v), the same functional relation must hold between the complex variables, so that we have

$$\psi + i\lambda = \tanh^{-1}\text{sn}\,(u + iv) - k\tanh^{-1}[k\,\text{sn}\,(u + iv)]. \tag{54.5}$$

If we use $(u/a, v/a)$ instead of (u, v), (54.5) is the inverse definition of a projection which we can call the Thompson projection. When $k = 0$, (54.5) reduces to

$$\tanh(\psi + i\lambda) = \sin(u + iv), \tag{54.6}$$

which, from (4.5), can be expressed as (12.3), the definition of the transverse Mercator projection of the sphere.

To separate the real and the imaginary parts of (54.5), we deal with the two terms separately. Thus, we can let

$$\tanh(\alpha + i\beta) = \text{sn}\,(u + iv), \tag{54.7}$$

from which we derive

$$\left.\begin{array}{l} \dfrac{\sinh 2\alpha}{\cosh 2\alpha + \cos 2\beta} = \dfrac{\text{sn}\,u\,\text{dn}'v}{1 - \text{dn}^2 u\,\text{sn}'^2 v} = g, \\[12pt] \dfrac{\sin 2\beta}{\cosh 2\alpha + \cos 2\beta} = \dfrac{\text{cn}\,u\,\text{dn}\,u\,\text{sn}'v\,\text{cn}'v}{1 - \text{dn}^2 u\,\text{sn}'^2 v} = h, \end{array}\right\} \tag{54.8}$$

whence

$$g^2 + h^2 = \frac{\cosh 2\alpha - \cos 2\beta}{\cosh 2\alpha + \cos 2\beta} = \frac{1 - cn^2 u \, cn'^2 v}{1 - dn^2 u \, sn'^2 v} . \qquad (54.9)$$

Thus we derive

$$\left. \begin{aligned} \tanh 2\alpha &= \frac{2g}{1 + g^2 + h^2} = \frac{2 \, sn \, u \, dn' v}{1 + sn^2 u \, dn'^2 v} , \\[2mm] \tan 2\beta &= \frac{2h}{1 - g^2 - h^2} = \frac{2 \, cn \, u \, dn \, u \, sn' v \, cn' v}{cn^2 u \, cn'^2 v - dn^2 u \, sn'^2 v} , \end{aligned} \right\} \qquad (54.10)$$

and finally

$$\tanh \alpha = sn \, u \, dn' v, \qquad \tan \beta = dc \, u \, sc' v. \qquad (54.11)$$

Similarly, for the second term of (54.5) we can let

$$\tanh (\alpha' + i\beta') = k \, sn \, (u + iv), \qquad (54.12)$$

from which we derive

$$\left. \begin{aligned} \frac{\sinh 2\alpha'}{\cosh 2\alpha' + \cos 2\beta'} &= \frac{k \, sn \, u \, dn' v}{1 - dn^2 u \, sn'^2 v} = g', \\[2mm] \frac{\sin 2\beta'}{\cosh 2\alpha' + \cos 2\beta'} &= \frac{k \, cn \, u \, dn \, u \, sn' v \, cn' v}{1 - dn^2 u \, sn'^2 v} = h', \end{aligned} \right\} \qquad (54.13)$$

whence

$$g'^2 + h'^2 = \frac{\cosh 2\alpha' - \cos 2\beta'}{\cosh 2\alpha' + \cos 2\beta'} = \frac{k^2 (1 - cn^2 u \, cn'^2 v)}{1 - dn^2 u \, sn'^2 v} . \qquad (54.14)$$

Thus we derive

$$\left. \begin{aligned} \tanh 2\alpha' &= \frac{2g'}{1 + g'^2 + h'^2} = \frac{2k \, sn \, u \, dn' v}{k^2 sn^2 u + dn'^2 v} , \\[2mm] \tan 2\beta' &= \frac{2h'}{1 - g'^2 - h'^2} = \frac{2k \, cn \, u \, dn \, u \, sn' v \, cn' v}{dn^2 u \, cn'^2 v - k^2 cn^2 u \, sn'^2 v} , \end{aligned} \right\} \qquad (54.15)$$

and finally

$$\tanh \alpha' = k \, sn \, u \, nd' v, \qquad \tan \beta' = k \, cd \, u \, sc' v. \qquad (54.16)$$

Using the results (54.11) and (54.16) in (54.5), we therefore have

$$\begin{aligned} \psi &= \tanh^{-1} (sn \, u \, dn' v) - k \tanh^{-1} (k \, sn \, u \, nd' v), \\ \lambda &= \tan^{-1} (dc \, u \, sc' v) - k \tan^{-1} (k \, cd \, u \, sc' v). \end{aligned} \qquad (54.17)$$

Writing (54.5) in the form

$$\zeta = \tanh^{-1} (sn \, w) - k \tanh^{-1} (k \, sn \, w), \qquad (54.18)$$

we get

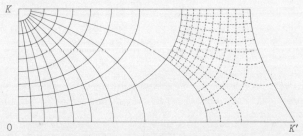

Fig. 43. Graticule of the Thompson transverse Mercator
projection of the International (Hayford) Spheroid.

$$\frac{d\zeta}{dw} = \frac{dn\,w}{cn\,w} - \frac{k^2\,cn\,w}{dn\,w} = \frac{k'^2}{cn\,w\,dn\,w}, \tag{54.19}$$

and therefore

$$\frac{dw}{d\zeta} = \frac{cn\,w\,dn\,w}{k'^2}. \tag{54.20}$$

By separating the real and the imaginary parts of this equation,
we obtain

$$\left.\begin{aligned}
\frac{\partial u}{\partial \psi} &= \frac{\partial v}{\partial \lambda} = \frac{cn\,u\,dn\,u\,dn'v\,(cn'^2 v - k^2 sn^2 u\,sn'^2 v)}{k'^2(1 - dn^2 u\,sn'^2 v)^2}, \\
-\frac{\partial v}{\partial \psi} &= \frac{\partial u}{\partial \lambda} = \frac{sn\,u\,sn'v\,cn'v\,(dn^2 u\,dn'^2 v + k^2 cn^2 u)}{k'^2(1 - dn^2 u\,sn'^2 v)^2}.
\end{aligned}\right\} \tag{54.21}$$

All four derivatives are zero at $u = 0$, $v = K'$, which is therefore
a singular point where there is a discontinuity in the equator.
The longitude at this point is $\lambda = \frac{1}{2}\pi(1 - k)$.

The graticule of one quadrant of a hemispheroid on the
Thompson projection is shown in Fig. 43, using a 10° interval in
latitude and longitude except in the outermost quadrilateral
where a 1° interval is used. The projection of the whole spheroid
is shown in Fig. 44. Both figures have been computed using the
eccentricity of the International (Hayford) Spheroid. On the
initial meridian, u is given by (54.1); the coordinates (u, v) in
other cases were computed by a method of successive approximations.

Scale and convergence can be obtained from (54.21). The
scale is given by

$$m = \frac{a}{\nu \cos \phi} \cdot \frac{(1 - sn^2 u\,dn'^2 v)^{1/2}\,(dn^2 u + dn'^2 v - 1)^{1/2}}{k'^2(1 - dn^2 u\,sn'^2 v)}. \tag{54.22}$$

On the initial meridian, from (54.2) and (54.3), $a/\nu \cos \phi = dc\,u$,
and (54.22) reduces to

$$m = dn^2 u/k'^2. \tag{54.23}$$

which ranges from 1 at the pole to $1/k'^2$ at the equator. On that
part of the equator for which $u = 0$,

$$m = dn'v/k'^2 cn'^2 v, \tag{54.24}$$

and on the meridian $\lambda = \frac{1}{2}\pi$, where $u = K$,

$$m = (a/\nu \cos \phi)\,sd'v\,cd'v. \tag{54.25}$$

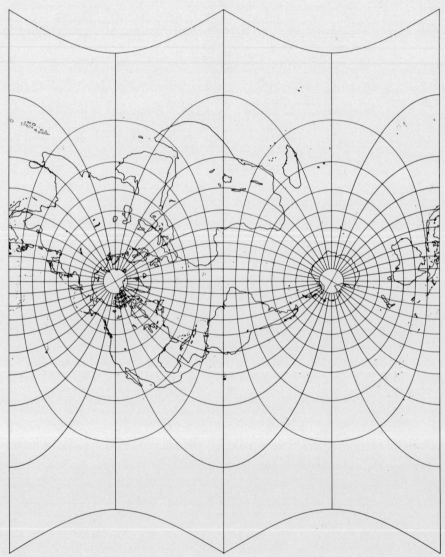

Fig. 44. Thompson transverse Mercator projection of the International (Hayford) Spheroid.

55. The Gauss-Krüger transverse Mercator projection of the
entire spheroid (Thompson 1945). The following definition,
suggested by Thomas Wray, is equivalent to the definition used by
Thompson. From (54.2) and (54.3), an element of arc of the meridian can be expressed as

$$dM = \rho\, d\phi = ak'^2\, \mathrm{nd}^2 u.$$ (55.1)

The length of a meridional arc is therefore given by

$$M = ak'^2 \int_0^u \mathrm{nd}^2 u\, du = a\, \mathrm{ed}\, u,$$ (55.2)

the constant of integration being zero since $M = 0$ when $u = 0$.

On the initial meridian of the Gauss-Krüger projection,
$x = M$, and therefore the projection coordinates $(x,\, y)$ can be
expressed in terms of the coordinates $(u,\, v)$ of the Thompson
projection (54.5) as

$$x + iy = a\, \mathrm{ed}\,(u + iv),$$ (55.3)

whence, by separating the real and the imaginary parts, we obtain
the projection coordinates as

$$\left.\begin{aligned}
\frac{x}{a} &= \mathrm{en}\, u - \frac{k^2 \mathrm{sn}\, u\, \mathrm{cn}\, u\, \mathrm{dn}\, u}{\mathrm{dn}^2 u + \mathrm{dn'}^2 v - 1}, \\[2mm]
\frac{y}{a} &= v - \mathrm{en'}v + \frac{k'^2 \mathrm{sn'}v\, \mathrm{cn'}v\, \mathrm{dn'}v}{\mathrm{dn}^2 u + \mathrm{dn'}^2 v - 1}.
\end{aligned}\right\}$$ (55.4)

For large values of v, the divisor in the last terms of (55.4)
becomes small, and extra decimals must be used to give adequate
accuracy in the quotient.

The graticule of the Gauss-Krüger projection, for one quadrant of a hemispheroid, as computed from (54.17) and (55.4), is
shown in Fig. 45. The x-axis represents the initial meridian
$\lambda = 0$, and on this meridian

$$x = a\, \mathrm{ed}\, u = M.$$ (55.5)

The line $x = aE$ represents the meridian $\lambda = \tfrac{1}{2}\pi$, and on this
meridian (where $u = K$)

$$y = a(v - \mathrm{en'}v + \mathrm{sc'}v\, \mathrm{dn'}v).$$ (55.6)

The positive y-axis represents part of the equator, extending from
$\lambda = 0$ to $\lambda = \tfrac{1}{2}\pi(1 - k)$, the latter value being at $y = a(K' - E')$.
At this point, the equator changes smoothly from a straight line
to a curve. On the y-axis, we have

$$y = a(v - \mathrm{ed'}v).$$ (55.7)

The projection of the entire spheroid is shown in Fig. 46,
again using the eccentricity of the International (Hayford) Spheroid. It can be seen that the entire spheroid is represented
within a finite area without singular points.

Writing (55.3) in the form

$$z = a\, \mathrm{ed}\, w,$$ (55.8)

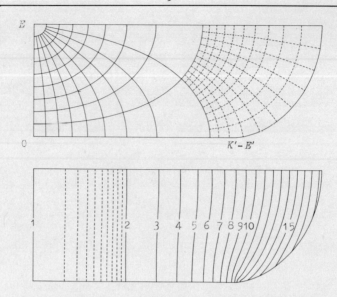

Fig. 45. Graticule and scale in the Gauss-Kruger transverse Mercator projection of the International (Hayford) Spheroid.

we have
$$d\mathit{s}/dw = ak'^2\,\mathrm{nd}^2 w,\qquad(55.9)$$

so that, with (54.20), we get

$$d\mathit{s}/d\zeta = a\,\mathrm{cd}\,w.\qquad(55.10)$$

Separation of the real and the imaginary parts gives

$$\frac{\partial x}{\partial \psi} = \frac{\partial y}{\partial \lambda} = \frac{a\,\mathrm{cn}\,u\,\mathrm{dn}\,u\,\mathrm{dn}'v}{\mathrm{dn}^2 u + \mathrm{dn}'^2 v - 1},$$

$$-\frac{\partial y}{\partial \psi} = \frac{\partial x}{\partial \lambda} = \frac{ak'^2\,\mathrm{sn}\,u\,\mathrm{sn}'v\,\mathrm{cn}'v}{\mathrm{dn}^2 u + \mathrm{dn}'^2 v - 1}.$$
$$\left.\right\} \quad (55.11)$$

The convergence is now given by

$$-\tan y = \frac{k'^2\,\mathrm{sn}\,u\,\mathrm{sn}'v\,\mathrm{cn}'v}{\mathrm{cn}\,u\,\mathrm{dn}\,u\,\mathrm{dn}'v} = \frac{\mathrm{dn}\,(K-u)}{\mathrm{sc}\,(K-u)}\cdot \mathrm{sn}'v\,\mathrm{sn}'(K'-v), \qquad (55.12)$$

and the scale coefficient is given by

$$m = \frac{a}{\nu \cos \phi}\left[\frac{1 - \mathrm{sn}^2 u\,\mathrm{dn}'^2 v}{\mathrm{dn}^2 u + \mathrm{dn}'^2 v - 1}\right]^{1/2}. \qquad (55.13)$$

On the initial meridian, $a/\nu \cos \phi = \mathrm{dc}\,u$, and (55.13) reduces to $m = 1$. On that part of the equator for which $u = 0$,

$$m = \mathrm{nd}'v, \qquad (55.14)$$

and on the meridian $\lambda = \tfrac{1}{2}\pi$, where $u = K$,

$$m = (a/\nu \cos \phi)\,\mathrm{sc}'v. \qquad (55.15)$$

Fig. 45 shows also the pattern of isomegeths.

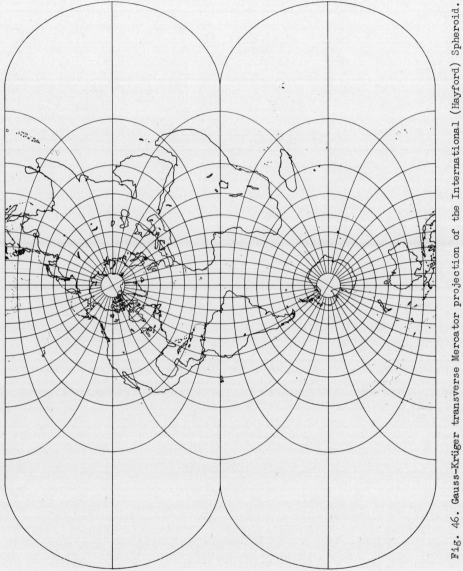

Fig. 46. Gauss-Krüger transverse Mercator projection of the International (Hayford) Spheroid.

56. Comparison and nomenclature of transverse Mercator projections of the spheroid. The three varieties of transverse Mercator projection described above can conveniently be compared, in the vicinity of the initial meridian, by means of series.

For the projection defined by (12.3), where ψ is the spheroidal isometric latitude (9.4), the definition $s = a \ \text{lam}^{-1} \zeta$ leads at once to

$$\frac{s}{a} = \zeta - \frac{1}{6}\zeta^3 + \frac{1}{24}\zeta^5 - \frac{61}{5040}\zeta^7 + \cdots , \qquad (56.1)$$

With s/a used instead of s, all the formulae of Sec. 12 apply to this projection, except that for scale coefficient. From (9.7) and (12.9) we get

$$m = \frac{a}{\nu \cos \phi} \ \frac{1}{(\cosh \psi - \sin \lambda)^{1/2}} \qquad (56.2)$$

On the initial meridian, this becomes $m = a/\nu \cos \phi \cosh \psi$, which varies from a minimum value of 1 at the equator to a maximum of $(1+k)^{\frac{1}{2}(1+k)} \ (1-k)^{\frac{1}{2}(1-k)}$ at the pole, the latter value being $1 \cdot 003\ 371$ for the Hayford spheroid.

For the Thompson projection, if ζ in (54.18) is expanded in powers of w and the series then reversed, we get

$$\frac{w}{a} = \frac{1}{1-k^2}\zeta - \frac{1+k^2}{6(1-k^2)^3}\zeta^3 + \frac{5 + 22k^2 + 5k^4}{120(1-k^2)^5}\zeta^5 -$$
$$- \frac{61 + 627k^2 + 627k^4 + 61k^6}{5040(1-k^2)^7}\zeta^7 + \cdots . \qquad (56.3)$$

As shown by (54.23), the scale on the initial meridian ranges from $1/k'^2$ at the equator to 1 at the pole, the scale at the equator being $1 \cdot 006\ 768$ for the Hayford spheroid.

For the Gauss-Krüger projection, with $m = 1$ along the initial meridian, we can expand $s = a \ \text{ed} \ w$ in powers of w, and substitute (56.3), with $a = 1$, in this series, to get

$$\frac{s}{a} = \zeta - \frac{1}{6(1-k^2)}\zeta^3 + \frac{5-k^2}{120(1-k^2)^3}\zeta^5 - \frac{61 + 26k^2 + k^4}{5040(1-k^2)^5}\zeta^7 + \cdots , \qquad (56.4)$$

which can be shown to be equivalent to the usual Taylor series for this projection.

It can be seen that, for $k = 0$, all three projections become identical with the transverse Mercator projection of the sphere.

A profusion of names has been applied to the first and third of these projections. In English-language publications, the name "transverse Mercator" is generally understood to refer to the projection given by (56.4), the constancy of scale along the initial meridian being regarded as the criterion for application of the name. A mathematician would be more inclined to apply it to the projection given by (56.1), since this is defined by the same mathematical function (12.3) as establishes the transverse Mercator projection of the sphere.

The projection with constant-scale initial meridian was originally due to Lambert 1772, who established it for the sphere and indicated how it could be modified for the spheroid. It was

next derived by Gauss c.1822, as one example of his investigations
in conformal representation, and was introduced by him into the
survey of Hannover. It is sometimes known as the Gauss conformal
projection, a description which is not sufficient to identify it,
and sometimes as Gauss's Hannover projection. Gauss left but few
details, and a more detailed analysis, aiming to reproduce Gauss's
method, was published by Schreiber 1866. Formulae adapted to
logarithmic computation were developed by Krüger 1912, whence came
the name, Gauss-Krüger projection, now used in many European
countries. Particular truncations of the defining series have
received other names, such as Gauss-Boaga as used in Italy.

The titles of the papers by Gauss, Schreiber, and Krüger,
with an indication of their content, are given in Jordan and
Eggert's *Handbuch der Vermessungskunde* (there is an English translation by the U. S. Army Map Service 1962), and from these it can
be seen that all three wrote about both projections. The name
Gauss-Krüger is therefore attached to the projection given by
(56.4) merely by usage, and not with complete historical justifi-
cation.

The projection given by (56.1) was due to Gauss 1843, who
derived it by first projecting the spheroid conformally upon a
sphere so that all the meridians correspond, and then making a
transverse Mercator projection of this sphere. It was described
by Schreiber 1897, and is sometimes known as Schreiber's double
projection. It was also described by Krüger 1914. The name
Gauss-Laborde is used to describe it in France. It was derived
in a more general form by Hotine 1947.

The names, Gauss-Schreiber and Gauss-Krüger, as used by Lee
1962 do not accord with European usage.

Conformal Projection of the Sphere upon a Regular Dodecahedron

57. Projection of the sphere upon a regular dodecahedron (Lee 1970). The foregoing work contains the development of conformal projections of the sphere upon four of the five regular polyhedra, so that a development of the fifth case is also of interest. Marvin 1929 expressed a desire to see the conformal projection of the sphere upon a regular dodecahedron developed, and supplied a sketch of it.

For the case $n = 5$, Schwarz's integral (3.1) cannot be identified with known functions, but the projection can be computed by series.

Derivation of the formula. The surface of the sphere can be divided into twelve equal spherical pentagons, each composed of ten spherical triangles, each with angles $\frac{1}{2}\pi$, $\frac{1}{3}\pi$, $\frac{1}{5}\pi$. The conformal projection of each such spherical triangle upon the infinite half-plane, with origin at a vertex with angle $\frac{1}{5}\pi$ and with the positive real axis passing through a vertex with angle $\frac{1}{3}\pi$, is given by (27.11), i.e.,

$$\left.\begin{aligned}
\xi &= \frac{1728r^5(1 + 11r^5 - r^{10})^5}{(1 + 522r^5 - 10005r^{10} - 10005r^{20} - 522r^{25} + r^{30})^2}, \\
1 - \xi &= \frac{(1 - 228r^5 + 494r^{10} + 228r^{15} + r^{20})^3}{(1 + 522r^5 - 10005r^{10} - 10005r^{20} - 522r^{25} + r^{30})^2}.
\end{aligned}\right\} \quad (57.1)$$

The conformal projection of a spherical pentagon within the unit circle, by (31.3), is given by

$$\begin{aligned}
z^5 &= (1 - \sqrt{1 - \xi})^2 / \xi \\
&= \frac{\begin{aligned}[1 + 522r^5 - 10005r^{10} - 10005r^{20} - 522r^{25} + r^{30} \\ - (1 - 228r^5 + 494r^{10} + 228r^{15} + r^{20})^{3/2}]^2\end{aligned}}{1728r^5(1 + 11r^5 - r^{10})^5} \\
&= \frac{432r^5(1 - 35r^5 - 760r^{10} - 67805r^{15} - \cdots)^2}{(1 + 11r^5 - r^{10})^5} \\
&= 432r^5(1 - 125r^5 + 5375r^{10} - 240500r^{15} - \cdots). \quad (57.2)
\end{aligned}$$

If we let

$$R^5 = 432r^5, \qquad \text{or} \qquad R = 3 \cdot 365\,865\,r, \quad (57.3)$$

(57.2) can be expressed as

$$z^5 = R^5 - \frac{125}{2^4 3^3}R^{10} + \frac{5375}{2^8 3^6}R^{15} - \frac{60125}{2^{10}3^9}R^{20} - \cdots, \quad (57.4)$$

whence we get

$$z = R - \frac{25}{2^4 3^3}R^6 - \frac{175}{2^8 3^6}R^{11} - \frac{17175}{2^{11}3^9}R^{16} - \cdots. \quad (57.5)$$

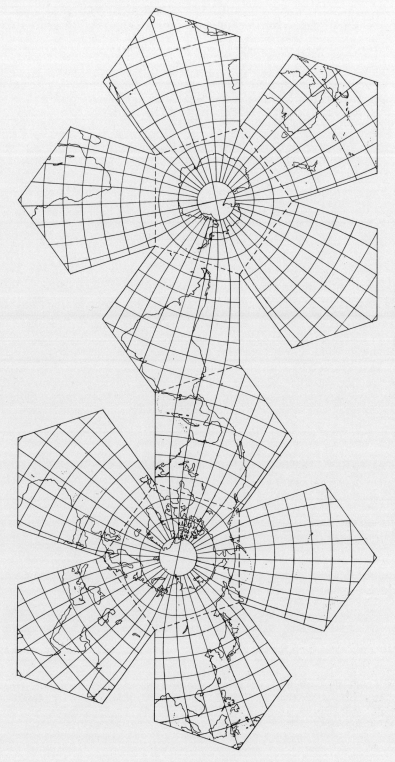

Fig. 47. Projection of the sphere upon a regular dodecahedron.

To represent the unit circle conformally within a regular pentagon, we have from Schwarz's integral (3.1)

$$w = \int_0^z (1 - z^5)^{-2/5}\, dz$$

$$= z + \frac{2}{5}\frac{1}{6}z^6 + \frac{2.7}{5.10}\frac{1}{11}z^{11} + \frac{2.7.12}{5.10.15}\frac{1}{16}z^{16} + \cdots$$

$$= R + \frac{19}{2160}R^6 + \frac{70243}{513\,21600}R^{11} + \frac{5\,77999}{16796\,16000}R^{16} + \cdots, \qquad (57.6)$$

or

$$\begin{aligned}
w = \quad & R & & +0\cdot000\,014\,R^{31}\\
& +0\cdot008\,796\,R^6 & & +0\cdot000\,005\,R^{36}\\
& +0\cdot001\,369\,R^{11} & & +0\cdot000\,002\,R^{41}\\
& +0\cdot000\,344\,R^{16} & & +0\cdot000\,001\,R^{46}\\
& +0\cdot000\,106\,R^{21} & & + \cdots.\\
& +0\cdot000\,037\,R^{26} & &
\end{aligned}$$

$$(57.7)$$

This series can be used for $|R| \not> 1$, but, as the maximum value of $|R|$ is $1\cdot138\,542$, it is not quite adequate for the whole of a pentagonal face, and many more terms would be needed to make it so.

The origin of coordinates is the centre of the pentagonal face and the positive real axis passes through a vertex.

The inverse series is

$$R = w - \frac{19}{2160}w^6 - \frac{46417}{513\,21600}w^{11} - \frac{12\,06709}{69284\,16000}w^{16} - \cdots, \qquad (57.8)$$

or

$$\begin{aligned}
R = \quad & w & & -0\cdot000\,012\,w^{26}\\
& -0\cdot008\,796\,w^6 & & -0\cdot000\,004\,w^{31}\\
& -0\cdot000\,904\,w^{11} & & -0\cdot000\,001\,w^{36}\\
& -0\cdot000\,174\,w^{16} & & - \cdots,\\
& -0\cdot000\,043\,w^{21} & &
\end{aligned}$$

$$(57.9)$$

which can be used for $|w| \not> 1$.

The circumradius of a pentagonal face is given by

$$K = \int_0^1 (1 - z^5)^{-2/5}\, dz = \tfrac{1}{5}\Gamma(\tfrac{1}{5})\,\Gamma(\tfrac{3}{5})/\Gamma(\tfrac{4}{5}) = 1\cdot174\,450. \quad (57.10)$$

and the length of a side is

$$K\sqrt{(5 - \sqrt{5})}/\sqrt{2} = 1\cdot380\,649. \qquad (57.11)$$

Nature of the projection. The sphere is conformally represented within twelve equal regular pentagons. In the example illustrated in Fig. 47, a geographic pole is the centre of two of the faces, and the other ten faces each have a centre given by $\tan \phi_0 = \pm\tfrac{1}{2}$, or $\phi_0 = \pm26° \,33' \,54''\cdot184$. The division of the world among the faces is the same as that sketched by Marvin and is arranged to show the Americas to best advantage.

Appendix I: Bibliography

ADAMS, Oscar S.—1925a: The rhombic conformal projection. *Bulletin Géodésique*, 1925: 1-26. Paris.

————1925b: *Elliptic Functions Applied to Conformal World Maps*. Special Publication No. 112, U.S. Coast and Geodetic Survey. Washington.

————1929: *Conformal Projection of the Sphere within a Square*. Special Publication No. 153, U.S. Coast and Geodetic Survey. Washington.

————1936: Conformal map of the world in a square, poles in the middle of opposite sides. *Bulletin Géodésique*, 1936: 461-473. Paris.

CAHILL, B.J.S.—1929: Projections for world maps. *Monthly Weather Review*, 57: 128-133. Washington.

COX, J.F.—1935: Représentation de la surface entière de la terre dans un triangle équilatéral. *Bulletin de la Classe des Sciences, Académie Royale de Belgique*, 5e, 21: 66-71. Brussels.

FORSYTH, A.R.—1893: *Theory of Functions of a Complex Variable*. University Press, Cambridge. [2nd ed., 1900].

GDOWSKI, Bogusław—1964: Odwzorowanie Gaussa-Krügera całej sferoidy. *Geodezjia i Kartografia*, 13: 209-229. Warsaw. Also *Survey Review*, 18 (1966): 339-341.

GOUGENHEIM, André—1950: Sur une nouvelle famille de planisphères conformes. *Annales Hydrographiques*, 4e, 1: 169-186. Paris

————1953: Emploi des projections conformes en cartographie. *Bulletin Géodésique*, 27: 7-38. Paris

GUYOU, Emile—1887: Nouveau système de projection de la sphère: Généralisation de la projection de Mercator. *Annales Hydrographiques*, 2e, 9: 16-35. Paris.

HOTINE, M.—1946-47: The orthomorphic projection of the spheroid. *Empire Survey Review*, 8:300-311; 9:25-35, 52-70, 112-123, 157-166. London.

KOBER, H.—1952: *Dictionary of Conformal Representations*. Dover Publications, Inc. New York.

LAMBERT, J.H.—1772: Anmerkungen und Zuzätse zur Entwerfung der Land- und Himmelscharten. Part III of *Beyträge zum Gebrauche der Matematik und deren Anwendung*, Verlag der Buchhandlung der Realschule, Berlin. Reprinted as No. 54 of Ostwald's *Klassiker der Exakten Wissenschaften*. English translation by W.R. Tobler, *Notes and Comments on the Composition of Terrestrial and Celestial Maps*, Michigan Geographical Publication No. 8, University of Michigan, Ann Arbor, 1972.

LEE, L.P.—1962: The transverse Mercator projection of the entire spheroid. *Empire Survey Review*, 16: 208-217. London. Also 17: 343.

————1963: Scale and convergence in the transverse Mercator projection of the entire spheroid. *Survey Review*, 17: 49-51. London.

————1965: Some conformal projections based on elliptic functions. *Geographical Review*, 55: 563-580. New York. Also 58 (1968): 490-491.

————1973: The conformal tetrahedric projection with some practical applications. *Cartographic Journal*, 10: 22-28. Edinburgh.

MAGIS, J.—1938: Calcul du canevas de la représentation conforme de la sphère entière dans un triangle équilatéral. *Bulletin Géodésique*, 1938: 247-256. Paris.

MARVIN, C.F.—1929: Projections for world maps. *Monthly Weather Review*, 57: 127. Washington.

PEIRCE, C.S.—1879: A quincuncial projection of the sphere. *American Journal of Mathematics*, 2: 394-396. Baltimore. Also *U.S. Coast Survey Report for the Year ending with June 1877*, 191-192. Washington.

SCHWARZ, H.A.—1869a: Ueber einige Abbildungsaufgaben. *Journal für die reine und angewandte Mathematik*, 70: 105-120. Berlin. *Gesammelte Mathematische Abhandlungen*, ii: 65-83.

————1869b: Conforme Abbildung der Oberfläche eines Tetraeders auf die Oberfläche einer Kugel. *Journal für die reine und angewandte Mathematik*, 70: 121-136. Berlin. *Gesammelte Mathematische Abhandlungen*, ii: 84-101.

————1869c: Notizia sulla rappresentazione conforme di un'area ellittica sopra un'area circolare. *Annali di Matematica pura ed applicata*, ii, 3: 166-170. Rome. *Gesammelte Mathematische Abhandlungen*, ii, 102-107.

————1872: Ueber diejenigen Fälle, in welchen die Gaussische hypergeometrische Reihe eine algebraische Function ihres vierten Elementes darstellt. *Journal für die reine und angewandte Mathematik*, 75: 292-335. Berlin. *Gesammelte Mathematische Abhandlungen*, ii: 211-259.

————1883: Zur conformen Abbildung der Fläche eines Rechtecks auf die Fläche einer Halbkugel. *Nachrichten von der Königlichen Gesellschaft der Wissenschaften und der Georg-Augusts-Universität zu Göttingen*, 1883: 51-60. *Gesammelte Mathematische Abhandlungen*, ii: 320-326.

WRAY, Thomas—1974: *The Seven Aspects of a General Map Projection. Cartographica* Monograph 11. B.V. Gutsell, York University, Toronto.

Appendix II: Mathematical Formulae

Series.

All the series used in this study have been found suitable for computation, without formal investigation of convergence.

Some special series used are:

$$(1 \pm x)^n = 1 \pm nx + \frac{n(n-1)}{2!} x^2 \pm \frac{n(n-1)(n-2)}{3!} x^3 + \cdots.$$

If n is a positive integer, the series ends at the term in x^n, and the coefficient of x^r is the binomial coefficient given by

$$\binom{n}{r} = \frac{n!}{(n-r)!\,r!}.$$

$$\int_0^z (1 - z^n)^{-2/n} \, dz$$

$$= z + \frac{2}{n} \frac{z^{n+1}}{n+1} + \frac{2(2+n)}{n \cdot 2n} \frac{z^{2n+1}}{2n+1} + \frac{2(2+n)(2+2n)}{n \cdot 2n \cdot 3n} \frac{z^{3n+1}}{3n+1} + \cdots.$$

Powers of series:

If $\quad y = x[1 + a_1 x^n + a_2 x^{2n} + a_3 x^{3n} + a_4 x^{4n} + \cdots]$,

then $\quad y^r = x^r \left[1 + \binom{r}{1} a_1 x^n + \left\{ \binom{r}{1} a_2 + \binom{r}{2} a_1^2 \right\} x^{2n} \right.$

$$+ \left\{ \binom{r}{1} a_3 + \binom{r}{2} 2 a_1 a_2 + \binom{r}{3} a_1^3 \right\} x^{3n}$$

$$\left. + \left\{ \binom{r}{1} a_4 + \binom{r}{2} (2 a_1 a_3 + a_2^2) + \binom{r}{3} a_1^2 a_2 + \binom{r}{4} a_1^4 \right\} x^{4n} + \cdots \right].$$

Formulae such as this have little practical value. In most cases, the powers can be formed more easily numerically than algebraically.

The rth root of a series can be formed by finding, term by term, the series which, when raised to the rth power, will give the original series.

Reversal of series:

If $\quad y = x[1 + a_1 x^n + a_2 x^{2n} + a_3 x^{3n} + \cdots]$,

then $\quad x = y[1 + b_1 y^n + b_2 y^{2n} + b_3 y^{3n} + \cdots]$,

where $\quad b_1 = -a_1$,

$$b_2 = (n+1)a_1^2 - a_2,$$

$$b_3 = -\tfrac{1}{2}(3n^2 + 5n + 2)a_1^3 + (3n+2)a_1 a_2 - a_3.$$

In practice, the b's can be found numerically by forming the powers of the first series, substituting these in the second series, and setting the coefficients of powers of x greater than the first equal to zero.

All reversals of series in this study have been checked by re-reversal.

Beta function (first Eulerian integral) in terms of Gamma functions

$$B(u, v) = \int_0^1 x^{u-1} (1 - x)^{v-1} dx = \Gamma(u)\,\Gamma(v)/\Gamma(u+v).$$

Functions of a complex variable

$$\text{lam } ix = i\,\text{lam}^{-1}x, \qquad \text{lam}^{-1}ix = i\,\text{lam } x.$$

$$(\cos\theta \pm i\sin\theta)^n = \cos n\theta \pm i\sin n\theta.$$

$$\exp(x \pm iy) = \exp x\,(\cos y \pm i\sin y).$$

$$\tan(x \pm iy) = \frac{\sin 2x \pm i\sinh 2y}{\cos 2x + \cosh 2y}.$$

$$\tanh(x \pm iy) = \frac{\sinh 2x \pm i\sin 2y}{\cosh 2x + \cos 2y}.$$

$$\tan^{-1}(x \pm iy) = \frac{1}{2}\left[\tan^{-1}\frac{2x}{1 - x^2 - y^2} \pm i\tanh^{-1}\frac{2y}{1 + x^2 + y^2}\right].$$

$$\tanh^{-1}(x \pm iy) = \frac{1}{2}\left[\tanh^{-1}\frac{2x}{1 + x^2 + y^2} \pm i\tan^{-1}\frac{2y}{1 - x^2 - y^2}\right].$$

The last two formulae above are usually avoided because of multivaluedness of the inverse tangent. This creates no difficulty in the use here made of these formulae, as any quadrant of the w-plane corresponds to the same quadrant of the z-plane.

Dixon elliptic functions for $\alpha = 0$.

A theory of elliptic functions, based on the curve, $x^3 + y^3 - 3\alpha xy = 1$, was developed by A. C. Dixon in 1890. Adams 1925b investigated more fully the case $\alpha = 0$. In the integral

$$u = \int_0^x (1 - x^3)^{-2/3}\,dx = \int_y^1 (1 - y^3)^{-2/3}\,dy,$$

put
$$x = \text{sm } u, \qquad y = \text{cm } u.$$

Then
$$\text{sm}^3 u + \text{cm}^3 u = 1.$$

$$\frac{d}{du}\text{sm } u = \text{cm}^2 u, \qquad \frac{d}{du}\text{cm } u = -\text{sm}^2 u.$$

K is the value of the integral when $x = 1$. The functions have a real period $3K$ and the complex periods $3\omega K$ and $3\omega^2 K$; only two of the three periods are independent.

The following summary contains the formulae which have been used in the foregoing study, nearly all of them given by Adams.

$$\text{sm } u = \text{cm}(K - u), \qquad\qquad \text{cm } u = \text{sm}(K - u),$$

$$\text{sm}(-u) = -\frac{\text{sm } u}{\text{cm } u}, \qquad\qquad \text{cm}(-u) = \frac{1}{\text{cm } u},$$

$$\text{sm}(K + u) = \frac{1}{\text{cm } u}, \qquad\qquad \text{cm}(K + u) = -\frac{\text{sm } u}{\text{cm } u},$$

$$\text{sm}(2K + u) = -\frac{\text{cm } u}{\text{sm } u}, \qquad\qquad \text{cm}(2K + u) = \frac{1}{\text{sm } u},$$

$$\text{sm}(3K + u) = \text{sm } u, \qquad\qquad \text{cm}(3K + u) = \text{cm } u.$$

$$\text{sm}\,\omega u = \omega\,\text{sm}\,u, \qquad \text{sm}\,\omega^2 u = \omega^2\text{sm}\,u., \qquad \text{cm}\,\omega u = \text{cm}\,\omega^2 u = \text{cm}\,u.$$

$$\text{sm}\,0 = \text{cm}\,K = 0, \qquad \text{sm}\,K = \text{cm}\,0 = 1, \qquad \text{sm}\tfrac{1}{2}K = \text{cm}\tfrac{1}{2}K = 2^{-1/3}.$$

$$\text{sm}\tfrac{1}{4}K = \text{cm}\tfrac{3}{4}K = \tfrac{1}{2}(1 + \sqrt{3} - \sqrt[4]{2\sqrt{3}}),$$

$$\text{sm}\tfrac{3}{4}K = \text{cm}\tfrac{1}{4}K = 2^{-5/3}\,(\sqrt{3}+1)(1 - \sqrt{3} + \sqrt[4]{2\sqrt{3}}).$$

See also some other special values in (32.2) above.

Addition formulae

$$\text{sm}\,(u+v) = \frac{\text{sm}^2 u\,\text{cm}\,v - \text{cm}\,u\,\text{sm}^2 v}{\text{sm}\,u\,\text{cm}^2 v - \text{cm}^2 u\,\text{sm}\,v} \qquad (u \neq v)$$

$$= \frac{\text{sm}\,u + \text{cm}^2 u\,\text{sm}\,v\,\text{cm}\,v}{\text{cm}\,v + \text{sm}\,u\,\text{cm}\,u\,\text{sm}^2 v} = \frac{\text{sm}\,v + \text{sm}\,u\,\text{cm}\,u\,\text{cm}^2 v}{\text{cm}\,u + \text{sm}^2 u\,\text{sm}\,v\,\text{cm}\,v}.$$

$$\text{cm}\,(u+v) = \frac{\text{sm}\,u\,\text{cm}\,u - \text{sm}\,v\,\text{cm}\,v}{\text{sm}\,u\,\text{cm}^2 v - \text{cm}^2 u\,\text{sm}\,v} \qquad (u \neq v)$$

$$= \frac{\text{cm}\,u\,\text{cm}^2 v - \text{sm}^2 u\,\text{sm}\,v}{\text{cm}\,v + \text{sm}\,u\,\text{cm}\,u\,\text{sm}^2 v} = \frac{\text{cm}^2 u\,\text{cm}\,v - \text{sm}\,u\,\text{sm}^2 v}{\text{cm}\,u + \text{sm}^2 u\,\text{sm}\,v\,\text{cm}\,v}.$$

$$\text{sm}\,2u = \frac{\text{sm}\,u\,(1 + \text{cm}^3 u)}{\text{cm}\,u\,(1 + \text{sm}^3 u)}, \qquad \text{cm}\,2u = \frac{\text{cm}^3 u - \text{sm}^3 u}{\text{cm}\,u\,(1 + \text{sm}^3 u)}.$$

Imaginary arguments

$$\text{sm}\,iv = \frac{i\sqrt{3}\,\text{sm}\,\dfrac{v}{\sqrt{3}}\,\text{cm}\,\dfrac{v}{\sqrt{3}}}{1 + \omega^2\text{sm}^3\,\dfrac{v}{\sqrt{3}}}, \qquad \text{cm}\,iv = \frac{1 + \omega\,\text{sm}^3\,\dfrac{v}{\sqrt{3}}}{1 + \omega^2\text{sm}^3\,\dfrac{v}{\sqrt{3}}}.$$

Series

$$\text{sm}^{-1} x = x F(\tfrac{2}{3},\ \tfrac{1}{3};\ \tfrac{4}{3};\ x^3)$$

$$= x + \frac{2}{3}\frac{1}{4}x^4 + \frac{2.5}{3.6}\frac{1}{7}x^7 + \frac{2.5.8}{3.6.9}\frac{1}{10}x^{10} + \cdots.$$

If $\quad \text{cm}\,u = \displaystyle\sum_{n=0}^{\infty} A_{3n}\,u^{3n}, \qquad \text{sm}\,u = \displaystyle\sum_{n=0}^{\infty} A_{3n+1}\,u^{3n+1},$

then the relations,

$$A_{3n} = -\frac{1}{3n}\sum_{r=0}^{n-1} A_{3r+1}\,A_{3n-3r-2}, \qquad A_{3n+1} = \frac{1}{3n+1}\sum_{r=0}^{n} A_{3r}\,A_{3n-3r},$$

provide a means of computing the successive coefficients, beginning with $A_0 = 1$.

$$\text{sm}\,u = u - \frac{1}{6}u^4 + \frac{2}{63}u^7 - \frac{13}{2268}u^{10} + \frac{23}{22113}u^{13} - \cdots,$$

$$\text{cm}\,u = 1 - \frac{1}{3}u^3 + \frac{1}{18}u^6 - \frac{23}{2268}u^9 + \frac{25}{13608}u^{12} - \cdots,$$

$$\text{sm}\,u\,\text{cm}\,u = u - \frac{1}{2}u^4 + \frac{1}{7}u^7 - \frac{1}{28}u^{10} + \frac{3}{364}u^{13} - \cdots.$$

Jacobian elliptic functions

In the elliptic integral of the first kind,

$$u = \int_0^\phi (1 - k^2\sin^2\phi)^{-1/2}\, d\phi, \qquad (k^2 < 1)$$

put $\qquad \phi = \operatorname{am} u, \qquad \sin\phi = \operatorname{sn} u, \qquad \cos\phi = \operatorname{cn} u,$

$(1 - k^2\sin^2\phi)^{1/2} = \operatorname{dn} u, \qquad d\phi = \operatorname{dn} u\, du, \qquad k'^2 = 1 - k^2.$

$$\operatorname{sn}^2 u + \operatorname{cn}^2 u = 1, \qquad \operatorname{dn}^2 u + k^2\operatorname{sn}^2 u = 1,$$
$$\operatorname{dn}^2 u - k^2\operatorname{cn}^2 u = k'^2, \qquad \operatorname{dn}^2 u - \operatorname{cn}^2 u = k'^2\operatorname{sn}^2 u.$$

In Glaisher's notation, the reciprocal of a Jacobian elliptic function is denoted by reversing the order of the letters ($1/\operatorname{sn} u = \operatorname{ns} u$), and a quotient by writing in order the initials of the numerator and the denominator ($\operatorname{sn} u/\operatorname{cn} u = \operatorname{sc} u$). In some textbooks $\operatorname{tn} u$ is used for $\operatorname{sc} u$. Glaisher's notation has the further property that, with appropriate lettering of the vertices of the period parallelogram, a function $\operatorname{pq} u$ has a zero at points p and a pole at points q.

$$\operatorname{sn}(-u) = -\operatorname{sn} u, \qquad \operatorname{cn}(-u) = \operatorname{cn} u, \qquad \operatorname{dn}(-u) = \operatorname{dn} u.$$

Hence, those functions involving the letter s change sign when u changes sign.

$$d\operatorname{sn} u/du = \operatorname{cn} u\operatorname{dn} u, \qquad\qquad d\operatorname{ns} u/du = -\operatorname{cs} u\operatorname{ds} u,$$
$$d\operatorname{cn} u/du = -\operatorname{sn} u\operatorname{dn} u, \qquad\qquad d\operatorname{cs} u/du = -\operatorname{ns} u\operatorname{ds} u,$$
$$d\operatorname{dn} u/du = -k^2\operatorname{sn} u\operatorname{cn} u, \qquad d\operatorname{ds} u/du = -\operatorname{ns} u\operatorname{cs} u,$$

$$d\operatorname{sc} u/du = \operatorname{nc} u\operatorname{dc} u, \qquad\qquad d\operatorname{sd} u/du = \operatorname{cd} u\operatorname{nd} u,$$
$$d\operatorname{nc} u/du = \operatorname{sc} u\operatorname{dc} u, \qquad\qquad d\operatorname{cd} u/du = -k'^2\operatorname{sd} u\operatorname{nd} u,$$
$$d\operatorname{dc} u/du = k'^2\operatorname{sc} u\operatorname{nc} u, \qquad\quad d\operatorname{nd} u/du = k^2\operatorname{sd} u\operatorname{cd} u.$$

K is the value of u when $\phi = \frac{1}{2}\pi$, and is a function of k. K' is the same function of k'.

The elliptic integral of the second kind is

$$\int_0^\phi (1 - k^2\sin^2\phi)^{1/2}d\phi = \int_0^u \operatorname{dn}^2 u\, du = \operatorname{en} u.$$

E is the value of this integral when $u = K$, and is a function of k. E' is the same function of k'.

$$EK' + E'K - KK' = \tfrac{1}{2}\pi.$$

We also define

$$k'^2 \int_0^u \operatorname{nd}^2 u = \operatorname{ed} u.$$

$$\operatorname{ed} u = \operatorname{en} u - k^2\operatorname{sn} u\operatorname{sn}(K - u).$$

The notations, $\operatorname{en} u$ and $\operatorname{ed} u$, were suggested by Thomas Wray and are part of his systematic notation for the integrals of the squares of elliptic functions which extends the zero-and-pole property of Glaisher's notation. Glaisher used $\operatorname{ez} u$ and $\operatorname{ez}_3 u$ respectively for these two functions. Adoption of the whole of Wray's notation involves using $gd\,u$ to denote one of these integrals instead of the gudermannian, and therefore we have used $lam^{-1}u$ to denote the gudermannian.

$$sn\,0 \;=\; 0, \qquad cn\,0 \;=\; 1, \qquad dn\,0 \;=\; 1, \qquad sc\,0 \;=\; 0,$$
$$sn\,K \;=\; 1, \qquad cn\,K \;=\; 0, \qquad dn\,K \;=\; k', \qquad sc\,K \;=\; \infty,$$
$$sn\tfrac{1}{2}K = \frac{1}{\sqrt{(1+k')}}, \qquad cn\tfrac{1}{2}K = \sqrt{\frac{k'}{1+k'}}, \qquad dn\tfrac{1}{2}K = \sqrt{k'} \qquad sc\tfrac{1}{2}K = \frac{1}{\sqrt{k'}}.$$
$$en\,0 \;=\; 0, \qquad en\,K \;=\; E, \qquad en\tfrac{1}{2}K \;=\; \tfrac{1}{2}(E+1-k'),$$
$$ed\,0 \;=\; 0, \qquad ed\,K \;=\; E, \qquad ed\tfrac{1}{2}K \;=\; \tfrac{1}{2}(E-1+k').$$

$$sn\,(K-u) \;=\; cd\,u \;=\; sn\,(K+u), \qquad cn\,(K-u) \;=\; k'sd\,u \;=\; -\,cn\,(K+u),$$
$$dn\,(K-u) \;=\; k'nd\,u \;=\; dn\,(K+u), \qquad sc\,(K-u) \;=\; (cs\,u)/k' \;=\; -\,sc\,(K+u).$$
$$en\,(K\pm u) \;=\; E\pm ed\,u, \qquad ed\,(K\pm u) \;=\; E\pm en\,u.$$

$$sn\tfrac{1}{2}u = \sqrt{\frac{1-cn\,u}{1+dn\,u}}, \qquad cn\tfrac{1}{2}u = \sqrt{\frac{cn\,u+dn\,u}{1+dn\,u}}, \qquad dn\tfrac{1}{2}u = \sqrt{\frac{cn\,u+dn\,u}{1+cn\,u}}.$$

$$sn\,2u = \frac{2\,sn\,u\,cn\,u\,dn\,u}{1-k^2\,sn^4u}, \qquad cn\,2u = \frac{cn^2u-sn^2u\,dn^2u}{1-k^2\,sn^4u},$$

$$dn\,2u = \frac{dn^2u-k^2\,sn^2u\,cn^2u}{1-k^2\,sn^4u}, \qquad sc\,2u = \frac{2\,sc\,u\,dn\,u}{1-sc^2u\,dn^2u}.$$

Series

The general terms of the series for Jacobian elliptic functions are not known, but there are relations between the coefficients from which the series can easily be formed. The series are

$$sn\,u = \sum_{n=0}^{\infty}(-1)^n S_{2n+1}\frac{u^{2n+1}}{(2n+1)!}, \qquad cn\,u = \sum_{n=0}^{\infty}(-1)^n C_{2n}\frac{u^{2n}}{(2n)!},$$

$$dn\,u = \sum_{n=0}^{\infty}(-1)^n D_{2n}\frac{u^{2n}}{(2n)!}, \qquad en\,u = \sum_{n=0}^{\infty}(-1)^n E_{2n+1}\frac{u^{2n+1}}{(2n+1)!},$$

where $C_0 = 1$, $D_0 = 1$, and

$$C_{2n} = \sum_{r=0}^{n-1}\binom{2n-1}{2r}D_{2r}S_{2n-2r-1}, \qquad D_{2n} = \sum_{r=0}^{n-1}\binom{2n-1}{2r}C_{2r}S_{2n-2r-1},$$

$$S_{2n+1} = \sum_{r=0}^{n}\binom{2n}{2r}D_{2r}C_{2n-2r}, \qquad E_{2n+1} = \sum_{r=0}^{n}\binom{2n}{2r}D_{2r}D_{2n-2r}.$$

$$sn\,u = u - (1+k^2)\frac{u^3}{3!} + (1+14k^2+k^4)\frac{u^5}{5!} - \cdots,$$

$$cn\,u = 1 - \frac{u^2}{2!} + (1+4k^2)\frac{u^4}{4!} - (1+44k^2+16k^4)\frac{u^6}{6!} + \cdots,$$

$$dn\,u = 1 - k^2\frac{u^2}{2!} + k^2(4+k^2)\frac{u^4}{4!} - k^2(16+44k^2+k^4)\frac{u^6}{6!} + \cdots,$$

$$en\,u = u - \tfrac{1}{3}k^2u^3 + \tfrac{1}{15}k^2(1+k^2)u^5 - \tfrac{1}{315}k^2(2+13k^2+2k^4)u^7 + \cdots.$$

$$ed\,u = (1-k^2)\left[u + \tfrac{1}{3}k^2u^3 - \tfrac{1}{15}k^2(1-2k^2)u^5 + \tfrac{1}{315}k^2(2-17k^2+17k^4)u^7 - \cdots\right].$$

Addition formulae

Let $\operatorname{sn} u = s_1$, $\operatorname{sn} v = s_2$, etc.   Then

$$\operatorname{sn}(u \pm v) = \frac{s_1 c_2 d_2 \pm c_1 d_1 s_2}{1 - k^2 s_1^2 s_2^2},$$

$$\operatorname{ns}(u \pm v) = \frac{s_1 c_2 d_2 \mp c_1 d_1 s_2}{s_1^2 - s_2^2},$$

$$\operatorname{cn}(u \pm v) = \frac{c_1 c_2 \mp s_1 d_1 s_2 d_2}{1 - k^2 s_1^2 s_2^2},$$

$$\operatorname{cs}(u \pm v) = \frac{s_1 c_1 d_2 \mp d_1 s_2 c_2}{s_1^2 - s_2^2},$$

$$\operatorname{dn}(u \pm v) = \frac{d_1 d_2 \mp k^2 s_1 c_1 s_2 c_2}{1 - k^2 s_1^2 s_2^2},$$

$$\operatorname{ds}(u \pm v) = \frac{s_1 d_1 c_2 \mp c_1 s_2 d_2}{s_1^2 - s_2^2},$$

$$\operatorname{sc}(u \pm v) = k \frac{s_1 c_1 d_2 \pm d_1 s_2 c_2}{d_1^2 d_2^2 - k'^2},$$

$$\operatorname{sd}(u \pm v) = \frac{s_1 d_1 c_2 \pm c_1 s_2 d_2}{k'^2 + k^2 c_1^2 c_2^2},$$

$$\operatorname{nc}(u \pm v) = k \frac{c_1 c_2 \pm s_1 d_1 s_2 d_2}{d_1^2 d_2^2 - k'^2},$$

$$\operatorname{cd}(u \pm v) = \frac{c_1 d_1 c_2 d_2 \mp k'^2 s_1 s_2}{k'^2 + k^2 c_1^2 c_2^2},$$

$$\operatorname{dc}(u \pm v) = k \frac{c_1 d_1 c_2 d_2 \pm k'^2 s_1 s_2}{d_1^2 d_2^2 - k'^2}$$

$$\operatorname{nd}(u \pm v) = \frac{d_1 d_2 \pm k^2 s_1 c_1 s_2 c_2}{k'^2 + k^2 c_1^2 c_2^2}.$$

$$
\begin{aligned}
\operatorname{en}(u \pm v) &= \operatorname{en} u \pm \operatorname{en} v \mp k^2 \operatorname{sn} u \operatorname{sn} v \operatorname{sn}(u \pm v) \\
&= \operatorname{ed} u \pm \operatorname{ed} v + k^2 \operatorname{cd} u \operatorname{cd} v \operatorname{sn}(u \pm v).
\end{aligned}
$$

$$
\begin{aligned}
\operatorname{ed}(u \pm v) &= \operatorname{en} u \pm \operatorname{en} v - k^2 \operatorname{cn} u \operatorname{cn} v \operatorname{sd}(u \pm v) \\
&= \operatorname{ed} u \pm \operatorname{ed} v \pm k^2 k'^2 \operatorname{sd} u \operatorname{sd} v \operatorname{sd}(u \pm v).
\end{aligned}
$$

Imaginary arguments

$$\operatorname{sn}(iu, k) = i \operatorname{sc}(u, k'), \qquad \operatorname{ns}(iu, k) = -i \operatorname{cs}(u, k'),$$
$$\operatorname{cn}(iu, k) = \operatorname{nc}(u, k'), \qquad \operatorname{cs}(iu, k) = -i \operatorname{ns}(u, k'),$$
$$\operatorname{dn}(iu, k) = \operatorname{dc}(u, k'), \qquad \operatorname{ds}(iu, k) = -i \operatorname{ds}(u, k'),$$

$$\operatorname{sc}(iu, k) = i \operatorname{sn}(u, k'), \qquad \operatorname{sd}(iu, k) = i \operatorname{sd}(u, k'),$$
$$\operatorname{nc}(iu, k) = \operatorname{cn}(u, k'), \qquad \operatorname{cd}(iu, k) = \operatorname{nd}(u, k'),$$
$$\operatorname{dc}(iu, k) = \operatorname{dn}(u, k'), \qquad \operatorname{nd}(iu, k) = \operatorname{cd}(u, k').$$

$$\operatorname{en}(iu, k) = i[u - \operatorname{en}(u, k') + \operatorname{sc}(u, k') \operatorname{dn}(u, k')].$$
$$\operatorname{ed}(iu, k) = i[u - \operatorname{ed}(u, k')].$$

Complex arguments

Let $\operatorname{sn}(u, k) = s$, $\operatorname{sn}(v, k') = s'$, etc.   Then

$$\operatorname{sn}(u \pm iv) = \frac{sd' \pm icds'c'}{1 - d^2 s'^2},$$

$$\operatorname{ns}(u \pm iv) = \frac{sd' \mp icds'c'}{1 - c^2 c'^2},$$

$$\operatorname{cn}(u \pm iv) = \frac{cc' \mp isds'd'}{1 - d^2 s'^2},$$

$$\operatorname{cs}(u \pm iv) = \frac{scc'd' \mp ids'}{1 - c^2 c'^2},$$

$$\operatorname{dn}(u \pm iv) = \frac{dc'd' \mp ik^2 scs'}{1 - d^2 s'^2},$$

$$\operatorname{ds}(u \pm iv) = \frac{sdc' \mp ics'd'}{1 - c^2 c'^2},$$

$$\text{sc}\,(u \pm iv) = \frac{scc'd' \pm ids'}{1 - s^2d'^2},$$
$$\text{sd}\,(u \pm iv) = \frac{sdc' \pm ics'd'}{d^2 + d'^2 - 1},$$

$$\text{nc}\,(u \pm iv) = \frac{cc' \pm isds'd'}{1 - s^2d'^2},$$
$$\text{cd}\,(u \pm iv) = \frac{cdd' \mp ik'^2 ss'c'}{d^2 + d'^2 - 1},$$

$$\text{dc}\,(u \pm iv) = \frac{cdd' \pm ik'^2 ss'c'}{1 - s^2d'^2},$$
$$\text{nd}\,(u \pm iv) = \frac{dc'd' \pm ik^2 scs'}{d^2 + d'^2 - 1}.$$

$$\text{en}\,(u \pm iv) = \text{en}\,u + \frac{k^2 scds'^2}{1 - d^2s'^2} \pm i\left[v - \text{en}'v + \frac{s'c'd'd^2}{1 - d^2s'^2}\right],$$

$$\text{ed}\,(u \pm iv) = \text{en}\,u - \frac{k^2 scd}{d^2 + d'^2 - 1} \pm i\left[v - \text{en}'v + \frac{k'^2 s'c'd'}{d^2 + d'^2 - 1}\right].$$

Identities associated with complex arguments

$$c'^2 + k^2s^2s'^2 = 1 - d^2s'^2, \qquad d^2d'^2 - k^2c^2 = k'^2(1 - d^2s'^2),$$
$$d'^2 - k^2c^2s'^2 = 1 - d^2s'^2, \qquad d^2 - k'^2s'^2 = d^2 + d'^2 - 1,$$
$$d^2c'^2 + k^2s^2 = 1 - d^2s'^2, \qquad c^2d'^2 + k'^2s^2c'^2 = d^2 + d'^2 - 1,$$
$$c^2c'^2 + s^2d'^2 = 1 - d^2s'^2, \qquad k^2c^2 + k'^2c'^2 = d^2 + d'^2 - 1.$$

$$s^2d'^2 + c^2d^2s'^2c'^2 = (1 - d^2s'^2)(1 - c^2c'^2),$$
$$s^2c^2c'^2d'^2 + d^2s'^2 = (1 - s^2d'^2)(1 - c^2c'^2),$$
$$c^2c'^2 + s^2d^2s'^2d'^2 = (1 - d^2s'^2)(1 - s^2d'^2),$$
$$s^2d^2c'^2 + c^2s'^2d'^2 = (1 - c^2c'^2)(d^2 + d'^2 - 1),$$
$$c^2d^2d'^2 + k'^4s^2s'^2c'^2 = (1 - s^2d'^2)(d^2 + d'^2 - 1),$$
$$d^2c'^2d'^2 + k^4s^2c^2s'^2 = (1 - d^2s'^2)(d^2 + d'^2 - 1).$$

$$k^2[1 - s^2d'^2 + k'(1 - c^2c'^2)] = (1 - k')(d^2 + k')(1 + k's'^2),$$
$$k^2[1 - s^2d'^2 - k'(1 - c^2c'^2)] = (1 + k')(d^2 - k')(1 - k's'^2),$$
$$1 - s^2d'^2 + k'^2(1 - c^2c'^2) = d^2 - d'^2 + 1,$$
$$(1 - s^2d'^2)(1 - d^2s'^2) + (1 - c^2c'^2)(d^2 + d'^2 - 1)$$
$$= (1 - k^2s^4)(1 - k'^2s'^4),$$
$$(1 - s^2d'^2)(1 - d^2s'^2) - (1 - c^2c'^2)(d^2 + d'^2 - 1)$$
$$= (c^2 - s^2d^2)(c'^2 - s'^2d'^2).$$

Miscellaneous identities

From a wealth of formulae relating to Jacobian elliptic functions, we have made use of the following.

$$\text{dn}\,2u - k' = \frac{(\text{dn}^2 u - k')^2}{(1 - k')(1 - k^2\text{sn}^4 u)},$$
$$\text{dn}\,2u + k' = \frac{(\text{dn}^2 u + k')^2}{(1 + k')(1 - k^2\text{sn}^4 u)},$$

$$1 + \text{dn}\,2u = \frac{2\,\text{dn}^2 u}{1 - k^2\text{sn}^4 u},$$
$$1 - \text{dn}\,2u = \frac{2k^2\text{sn}^2 u\,\text{cn}^2 u}{1 - k^2\text{sn}^4 u},$$

$$\frac{1 + \text{dn}\,2u}{\text{dn}\,2u - k'} = \frac{2(1 - k')\,\text{dn}^2 u}{(\text{dn}^2 u - k')^2},$$
$$\frac{1 - \text{dn}\,2u}{\text{dn}\,2u + k'} = \frac{2k^2(1 + k')\,\text{sn}^2 u\,\text{cn}^2 u}{(\text{dn}^2 u + k')^2}.$$

$$1 - k \operatorname{sn} (K - 2u) = \frac{(1 - k)(1 + k \operatorname{sn}^2 u)^2}{\operatorname{dn}^2 u - k^2 \operatorname{sn}^2 u \operatorname{cn}^2 u},$$

$$1 + k \operatorname{sn} (K - 2u) = \frac{(1 + k)(1 - k \operatorname{sn}^2 u)^2}{\operatorname{dn}^2 u - k^2 \operatorname{sn}^2 u \operatorname{cn}^2 u},$$

$$1 - \operatorname{sn} (K - 2u) = \frac{2 k'^2 \operatorname{sn}^2 u}{\operatorname{dn}^2 u - k^2 \operatorname{sn}^2 u \operatorname{cn}^2 u},$$

$$1 + \operatorname{sn} (K - 2u) = \frac{2 \operatorname{cn}^2 u \operatorname{dn}^2 u}{\operatorname{dn}^2 u - k^2 \operatorname{sn}^2 u \operatorname{cn}^2 u},$$

$$\frac{1 - \operatorname{sn} (K - 2u)}{1 + k \operatorname{sn} (K - 2u)} = \frac{2 k'^2 \operatorname{sn}^2 u}{(1 + k)(1 - k \operatorname{sn}^2 u)^2},$$

$$\frac{1 + \operatorname{sn} (K - 2u)}{1 - k \operatorname{sn} (K - 2u)} = \frac{2 \operatorname{cn}^2 u \operatorname{dn}^2 u}{(1 - k)(1 + k \operatorname{sn}^2 u)^2},$$

$$\frac{\operatorname{sn} u \operatorname{cn} u}{\operatorname{dn} u} = \frac{1}{k} \sqrt{\frac{1 - \operatorname{dn} 2u}{1 + \operatorname{dn} 2u}},$$

$$\frac{\operatorname{cn} u \operatorname{dn} u}{\operatorname{sn} u} = k' \sqrt{\frac{1 + \operatorname{sn} (K - 2u)}{1 - \operatorname{sn} (K - 2u)}}.$$

Limiting values of modulus

For $k = 0$, $k' = 1$:
$K = \frac{1}{2}\pi$, $K' = \infty$, $E = \frac{1}{2}\pi$, $E' = 1$, $\operatorname{am} u = u$, $\operatorname{sn} u = \sin u$, $\operatorname{cn} u = \cos u$, $\operatorname{dn} u = 1$, $\operatorname{scu} = \tan u$, $\operatorname{en} u = \operatorname{ed} u = u$.

For $k = 1$, $k' = 0$:
$K = \infty$, $K' = \frac{1}{2}\pi$, $E = 1$, $E' = \frac{1}{2}\pi$, $\operatorname{am} u = \operatorname{lam}^{-1} u$, $\operatorname{sn} u = \operatorname{scu} = \tanh u$, $\operatorname{cn} u = \operatorname{dn} u = \operatorname{sech} u$, $\operatorname{en} u = \sinh u$, $\operatorname{ed} u = 0$ $(u \neq K)$.

Appendix III: Mathematical Tables

Latitude functions for the sphere.

Table 1 gives, for every degree of latitude ϕ, the value of ϕ in radians, the isometric latitude ψ, and functions of $\frac{1}{2}\psi$. Functions of ψ can be obtained from trigonometric tables, as shown by (4.5) above.

Dixon elliptic functions for $\alpha = 0$.

Adams published a 4-decimal table of $\operatorname{sm} u$ and $\operatorname{cm} u$ at interval $K/120$ from $u = 0$ to $u = K$. A 12-decimal table at interval $K/100$ is given here (Table 2). Derived from it is the 5-decimal table at interval $K/10$ for the complete range from $-3K$ to $+3K$ (Table 3).

Tables 4 to 10 are for use with the methods described in Sec. 32 above for separating the real and the imaginary parts of $\operatorname{sm} w$, that is, for finding u from $\operatorname{sm} 2u$ and finding v from $\operatorname{sm}^3(2v/\sqrt{3})$, and cover the ranges, from $-\frac{1}{2}K$ to $+K$ of u and from $-\frac{1}{2}\sqrt{3}K$ to $+\frac{1}{2}\sqrt{3}K$ of v, required in the transformation $\operatorname{sm} w = z$.

To keep the tables within reasonable compass and to give 6-decimal accuracy without using differences higher than the third, it is necessary to adopt different procedures in different parts of the range, as follows.

Range of $\operatorname{sm} 2u$	Procedure
0 to 0·995	Table 4
0·995 to 1·005	Table 5 (inverse)
> 1·005	Use reciprocal and Table 4 or 5
0 to −1	Table 6
< −1	Use reciprocal and Table 6

Range of $\operatorname{sm}^3(2v/\sqrt{3})$	Procedure
0 to 0·05	Table 7 (inverse)
0·05 to 0·95	Table 9
0·95 to 1·00	Use complement and Table 7
0 to −0·05	Table 8 (inverse)
−0.05 to −1	Table 10
< −1	Use reciprocal and Table 10

An accurate value of u cannot be obtained from $\operatorname{sm} 2u$ when $\operatorname{sm} 2u$ is very close to 1, and an accurate value of v cannot be obtained from $\operatorname{sm}^3(2v/\sqrt{3})$ when v is very small.

Differences are for use with the Bessel central difference interpolation formula. $2M^2$ denotes the "double second difference" $\delta_0^2 + \delta_1^2$, modified for the throwback from fourth differences.

Jacobian elliptic functions for $k = \sin 45°$.

Tables 11-14, for finding u from $\operatorname{cn} 2\sqrt{2}u$ for $k = \sin 45°$, are based on the *Smithsonian Elliptic Functions Tables* by G. W. and R. M. Spenceley, Smithsonian Institution, Washington, 1947.

Tables of $\operatorname{sn} u$, $\operatorname{cn} u$, $\operatorname{dn} u$, $\operatorname{en} u$, for the International Spheroid ($k^2 = 593/88209$) were also computed, but are not thought worth reproducing here.

TABLE 1. LATITUDE FUNCTIONS FOR THE SPHERE

φ°	arc φ	ψ	sinh ½ψ	cosh ½ψ	tanh ½ψ
0	0·000 000	0·000 000	0·000 000	1·000 000	0·000 000
1	·017 453	·017 454	·008 727	1·000 038	·008 727
2	·034 907	·034 914	·017 458	1·000 152	·017 455
3	·052 360	·052 384	·026 195	1·000 343	·026 186
4	·069 813	·069 870	·034 942	1·000 610	·034 921
5	0·087 266	0·087 377	0·043 703	1·000 955	0·043 661
6	·104 720	·104 912	·052 480	1·001 376	·052 408
7	·122 173	·122 478	·061 277	1·001 876	·061 163
8	·139 626	·140 082	·070 098	1·002 454	·069 927
9	·157 080	·157 730	·078 947	1·003 111	·078 702
10	0·174 533	0·175 426	0·087 825	1·003 849	0·087 489
11	·191 986	·193 177	·096 739	1·004 668	·096 289
12	·209 440	·210 988	·105 690	1·005 570	·105 104
13	·226 893	·228 865	·114 682	1·006 555	·113 936
14	·244 346	·246 814	·123 721	1·007 624	·122 785
15	0·261 799	0·264 842	0·132 808	1·008 780	0·131 652
16	·279 253	·282 955	·141 950	1·010 025	·140 541
17	·296 706	·301 158	·151 149	1·011 358	·149 451
18	·314 159	·319 458	·160 409	1·012 784	·158 384
19	·331 613	·337 863	·169 736	1·014 303	·167 343
20	0·349 066	0·356 379	0·179 134	1·015 918	0·176 327
21	·366 519	·375 012	·188 607	1·017 631	·185 339
22	·383 972	·393 771	·198 160	1·019 445	·194 380
23	·401 426	·412 663	·207 798	1·021 362	·203 452
24	·418 879	·431 695	·217 527	1·023 386	·212 557
25	0·436 332	0·450 875	0·227 352	1·025 519	0·221 695
26	·453 786	·470 213	·237 278	1·027 765	·230 868
27	·471 239	·489 715	·247 312	1·030 128	·240 079
28	·488 692	·509 392	·257 459	1·032 611	·249 328
29	·506 145	·529 253	·267 726	1·035 218	·258 618
30	0·523 599	0·549 306	0·278 119	1·037 955	0·267 949
31	·541 052	·569 563	·288 646	1·040 825	·277 325
32	·558 505	·590 033	·299 315	1·043 834	·286 745
33	·575 959	·610 728	·310 132	1·046 987	·296 213
34	·593 412	·631 658	·321 106	1·050 290	·305 731
35	0·610 865	0·652 837	0·332 246	1·053 749	0·315 299
36	·628 319	·674 275	·343 561	1·057 371	·324 920
37	·645 772	·695 988	·355 060	1·061 163	·334 595
38	·663 225	·717 988	·366 755	1·065 133	·344 328
39	·680 678	·740 290	·378 655	1·069 289	·354 119
40	0·698 132	0·762 910	0·390 773	1·073 640	0·363 970
41	·715 585	·785 863	·403 121	1·078 196	·373 885
42	·733 038	·809 167	·415 712	1·082 966	·383 864
43	·750 492	·832 841	·428 560	1·087 963	·393 910
44	·767 945	·856 903	·441 681	1·093 198	·404 026
45	0·785 398	0·881 374	0·455 090	1·098 684	0·414 214

TABLE 1. LATITUDE FUNCTIONS FOR THE SPHERE

ϕ	arc ϕ	ψ	sinh $\frac{1}{2}\psi$	cosh $\frac{1}{2}\psi$	tanh $\frac{1}{2}\psi$
45	0·785 398	0·881 374	0·455 090	1·098 684	0·414 214
46	·802 851	·906 275	·468 805	1·104 436	·424 475
47	·820 305	·931 632	·482 845	1·110 468	·434 812
48	·837 758	·957 467	·497 231	1·116 798	·445 229
49	·855 211	0·983 808	·511 983	1·123 444	·455 726
50	0·872 665	1·010 683	0·527 126	1·130 426	0·466 308
51	·890 118	1·038 123	·542 686	1·137 764	·476 976
52	·907 571	1·066 162	·558 690	1·145 484	·487 733
53	·925 025	1·094 833	·575 170	1·153 612	·498 582
54	·942 478	1·124 177	·592 158	1·162 175	·509 525
55	0·959 931	1·154 235	0·609 691	1·171 206	C·520 567
56	0·977 384	1·185 051	·627 810	1·180 740	·531 709
57	0·994 838	1·216 675	·646 560	1·190 815	·542 956
58	1·012 291	1·249 161	·665 988	1·201 474	·554 309
59	1·029 744	1·282 567	·686 150	1·212 766	·565 773
60	1·047 198	1·316 958	0·707 107	1·224 745	0·577 350
61	1·064 651	1·352 405	·728 926	1·237 470	·589 045
62	1·082 104	1·388 986	·751 683	1·251 010	·600 861
63	1·099 557	1·426 788	·775 464	1·265 442	·612 801
64	1·117 011	1·465 908	·800 366	1·280 854	·624 869
65	1·134 464	1·506 454	0·826 499	1·297 344	0·637 070
66	1·151 917	1·548 547	·853 989	1·315 027	·649 408
67	1·169 371	1·592 324	·882 979	1·334 036	·661 886
68	1·186 824	1·637 939	·913 638	1·354 523	·674 509
69	1·204 277	1·685 568	·946 158	1·376 668	·687 281
70	1·221 730	1·735 415	0·980 766	1·400 679	0·700 208
71	1·239 184	1·787 712	1·017 731	1·426 806	·713 293
72	1·256 637	1·842 730	1·057 371	1·455 347	·726 543
73	1·274 090	1·900 787	1·100 069	1·486 658	·739 961
74	1·291 544	1·962 257	1·146 289	1·521 176	·753 554
75	1·308 997	2·027 589	1·196 600	1·559 440	0·767 327
76	1·326 450	2·097 324	1·251 712	1·602 118	·781 286
77	1·343 904	2·172 122	1·312 519	1·650 062	·795 436
78	1·361 357	2·252 803	1·380 169	1·704 367	·809 784
79	1·378 810	2·340 401	1·456 167	1·766 471	·824 336
80	1·396 263	2·436 246	1·542 526	1·838 310	0·839 100
81	1·413 717	2·542 090	1·642 019	1·922 557	·854 081
82	1·431 170	2·660 306	1·758 593	2·023 029	·869 287
83	1·448 623	2·794 219	1·898 092	2·145 403	·884 725
84	1·466 077	2·948 700	2·069 634	2·298 562	·900 404
85	1·483 530	3·131 301	2·288 418	2·497 370	0·916 331
86	1·500 983	3·354 674	2·582 207	2·769 078	·932 515
87	1·518 436	3·642 533	3·008 930	3·170 751	·948 965
88	1·535 890	4·048 125	3·718 448	3·850 565	·965 689
89	1·553 343	4·741 349	5·305 596	5·399 013	0·982 697
90	1·570 796	∞	∞	∞	1·000 000

TABLE 2.   DIXON ELLIPTIC FUNCTIONS for α = 0

K = 1·76663 87502 85   1/K = 0·56604 66803 63

u/K	sm u	cm u	
0·00	0·00000 00000 00	1·00000 00000 00	1·00
·01	·01766 63712 68	0·99999 81621 01	0·99
·02	·03533 25152 55	·99998 52969 04	·98
·03	·05299 78475 49	·99995 03779 17	·97
·04	·07066 13942 51	·99988 23812 84	·96
0·05	0·08832 17922 64	0·99977 02888 21	0·95
·06	·10597 72898 36	·99960 30922 68	·94
·07	·12362 57474 47	·99936 97987 53	·93
·08	·14126 46391 20	·99905 94374 48	·92
·09	·15889 10542 55	·99866 10674 30	·91
0·10	0·17650 17000 44	0·99816 37867 15	0·90
·11	·19409 29045 73	·99755 67424 56	·89
·12	·21166 06206 73	·99682 91422 81	·88
·13	·22920 04306 10	·99597 02667 45	·87
·14	·24670 75516 81	·99496 94828 61	·86
0·15	0·26417 68427 82	0·99381 62586 70	0·85
·16	·28160 28120 24	·99250 01788 07	·84
·17	·29897 96254 60	·99101 09610 05	·83
·18	·31630 11169 62	·98933 84734 81	·82
·19	·33356 07993 14	·98747 27531 33	·81
0·20	0·35075 18765 62	0·98540 40244 71	0·80
·21	·36786 72576 43	·98312 27192 00	·79
·22	·38489 95713 33	·98061 94963 66	·78
·23	·40184 11825 31	·97788 52529 62	·77
·24	·41868 42098 72	·97491 11948 82	·76
0·25	0·43542 05446 82	0·97168 87581 28	0·75
·26	·45204 18712 61	·96820 97301 24	·74
·27	·46853 96884 51	·96446 62210 34	·73
·28	·48490 53324 80	·96045 06949 40	·72
·29	·50113 00010 12	·95615 59907 38	·71
0·30	0·51720 47783 52	0·95157 53426 34	0·70
·31	·53312 06617 34	·94670 24000 71	·69
·32	·54886 85886 08	·94153 12469 60	·68
·33	·56443 94648 23	·93605 64200 71	·67
·34	·57982 41936 17	·93027 29264 20	·66
0·35	0·59501 37052 73	0·92417 62595 47	0·65
·36	·60999 89873 32	·91776 24145 03	·64
·37	·62477 11152 22	·91102 79014 53	·63
·38	·63932 12831 51	·90396 97577 45	·62
·39	·65364 08351 23	·89658 55583 28	·61
0·40	0·66772 12959 10	0·88887 34244 24	0·60
·41	·68155 44018 26	·88083 20303 35	·59
·42	·69513 21311 30	·87246 06083 12	·58
·43	·70844 67339 02	·86375 89514 04	·57
·44	·72149 07612 15	·85472 74142 29	·56
0·45	0·73425 70934 45	0·84536 69116 20	0·55
·46	·74673 89675 53	·83567 89151 05	·54
·47	·75893 00031 92	·82566 54472 30	·53
·48	·77082 42274 79	·81532 90736 92	·52
·49	·78241 60982 95	·80467 28933 29	·51
0·50	0·79370 05259 84	0·79370 05259 84	0·50
	cm u	sm u	u/K

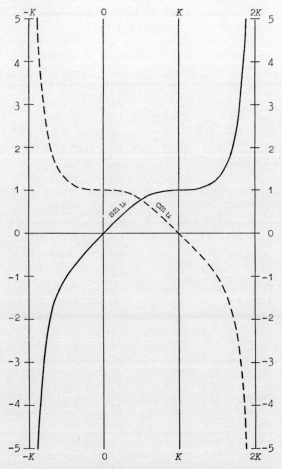

Fig. 48. Graph of sm u and cm u.

TABLE 3. $u = -3K(K/10)+3K$.

u/K	u/K	sm u	cm u		
-2·5	0·5	0·79370	0·79370	0·5	-2·5
-2·4	0·6	0·88887	0·66772	0·4	-2·6
-2·3	0·7	0·95158	0·51720	0·3	-2·7
-2·2	0·8	0·98540	0·35075	0·2	-2·8
-2·1	0·9	0·99816	0·17650	0·1	-2·9
-2·0	1·0	1·00000	0·00000	0·0	±3·0
-1·9	1·1	1·00184	-0·17683	-0·1	2·9
-1·8	1·2	1·01481	-0·35595	-0·2	2·8
-1·7	1·3	1·05089	-0·54352	-0·3	2·7
-1·6	1·4	1·12502	-0·75120	-0·4	2·6
-1·5	1·5	1·25992	-1·00000	-0·5	2·5
-1·4	1·6	1·49763	-1·33120	-0·6	2·4
-1·3	1·7	1·93347	-1·83984	-0·7	2·3
-1·2	1·8	2·85102	-2·80940	-0·8	2·2
-1·1	1·9	5·66567	-5·65526	-0·9	2·1
-1·0	2·0	±∞	±∞	-1·0	2·0
		cm u	sm u	u/K	u/K

TABLES FOR FINDING u FROM sm $2u$

TABLE 4

sm $2u$	u	δ^1 +	$2\delta^2$ +		sm $2u$	u	δ^1 +	$2M^2$ +	δ^3 +
0·00	0·000 000				0·50	0·255 545			
		5000					5481	64	
·01	·005 000	5000			·51	·261 026	5515	68	
·02	·010 000	5000			·52	·266 541	5549	71	
·03	·015 000	5000			·53	·272 090	5586	76	
·04	·020 000	5001			·54	·277 676	5625	79	
0·05	0·025 001				0·55	0·283 301			
		5000					5665	84	
·06	·030 001	5001			·56	·288 966	5709	90	
·07	·035 002	5001			·57	·294 675	5755	94	
·08	·040 003	5002			·58	·300 430	5803	99	
·09	·045 005	5003			·59	·306 233	5854	106	
0·10	0·050 008				0·60	0·312 087			
		5004					5909	111	
·11	·055 012	5005			·61	·317 996	5965	117	
·12	·060 017	5007			·62	·323 961	6026	125	
·13	·065 024	5008			·63	·329 987	6090	132	
·14	·070 032	5010			·64	·336 077	6158	140	
0·15	0·075 042				0·65	0·342 235			
		5013					6230	149	
·16	·080 055	5015			·66	·348 465	6307	158	
·17	·085 070	5018			·67	·354 772	6388	168	
·18	·090 088	5021	7		·68	·361 160	6475	180	
·19	·095 109	5025	8		·69	·367 635	6568	192	
0·20	0·100 134		8		0·70	0·374 203			
		5029					6667	204	
·21	·105 163	5033	10		·71	·380 870	6772	217	
·22	·110 196	5039	10		·72	·387 642	6884	235	
·23	·115 235	5043	11		·73	·394 526	7007	252	
·24	·120 278	5050	13		·74	·401 533	7136	270	
0·25	0·125 328		13		0·75	0·408 669			
		5056					7277	292	
·26	·130 384	5063	15		·76	·415 946	7428	315	
·27	·135 447	5071	15		·77	·423 374	7592	342	
·28	·140 518	5078	17		·78	·430 966	7770	372	
·29	·145 596	5088	19		·79	·438 736	7964	406	
0·30	0·150 684		19		0·80	0·446 700			
		5097					8176	444	
·31	·155 781	5107	21		·81	·454 876	8408	488	
·32	·160 888	5118	22		·82	·463 284	8664	540	
·33	·166 006	5129	24		·83	·471 948	8948	599	
·34	·171 135	5142	26		·84	·480 896	9263	669	
0·35	0·176 277		27		0·85	0·490 159			
		5155					9617	754	55
·36	·181 432	5169	29		·86	·499 776	10017	851	67
·37	·186 601	5184	30		·87	·509 793	10472	971	86
·38	·191 785	5199	33		·88	·520 265	10994	1122	108
·39	·196 984	5217	36		·89	·531 259	11602	1314	
0·40	0·202 201		36		0·90	0·542 861			
		5235					12318	1559	143
·41	·207 436	5253	39		·91	·555 179	13177	1888	193
·42	·212 689	5274	41		·92	·568 356	14229	2336	270
·43	·217 963	5294	43		·93	·582 585	15551	2975	400
·44	·223 257	5317	47		·94	·598 136	17273	3943	633
0·45	0·228 574		49		0·95	0·615 409			
		5341							
·46	·233 915	5366	52						
·47	·239 281	5393	54						
·48	·244 674	5420	59						
·49	·250 094	5451	61						
0·50	0·255 545								

For continuation
see next page.

For values of sm $2u$ greater than 1·005, evaluate the reciprocal. Enter Table 4 with this reciprocal as argument, and subtract the derived value of u from 1·766 639. For example, if sm $2u$ = 1·386 274, the reciprocal is 0·721 358. With this value as argument, we find the table gives u = 0·388 570. The required value is therefore u = 1·378 069.

TABLES FOR FINDING u FROM sm 2u

TABLE 4 (continued)

sm 2u	u	δ¹	2M²	δ³
0·950	0·615 409	+	+	+
		1843	48	
·951	·617 252	1868	49	
·952	·619 120	1892	50	
·953	·621 012	1918	52	
·954	·622 930	1944	54	
0·955	0·624 874			
		1972	58	
·956	·626 846	2002	58	
·957	·628 848	2030	61	
·958	·630 878	2063	64	
·959	·632 941	2094	65	
0·960	0·635 035			
		2128	70	
·961	·637 163	2164	72	
·962	·639 327	2200	75	
·963	·641 527	2239	79	
·964	·643 766	2279	82	
0·965	0·646 045			
		2321	87	
·966	·648 366	2366	91	
·967	·650 732	2412	96	
·968	·653 144	2462	101	
·969	·655 606	2513	106	
0·970	0·658 119			
		2568	113	
·971	·660 687	2626	120	
·972	·663 313	2688	127	
·973	·666 001	2753	134	
·974	·668 754	2822	144	
0·975	0·671 576			
		2897	155	
·976	·674 473	2977	165	
·977	·677 450	3062	177	
·978	·680 512	3154	192	
·979	·683 666	3254	208	
0·980	0·686 920			
		3362	225	
·981	·690 282	3479	247	
·982	·693 761	3609	272	
·983	·697 370	3751	298	
·984	·701 121	3907	332	
0·985	0·705 028			
		4083	372	
·986	·709 111	4279	419	
·987	·713 390	4502	475	
·988	·717 892	4756	544	
·989	·722 648	5050	635	52
0·990	0·727 698			
		5396	751	66
·991	·733 094	5808	903	90
·992	·738 902	6310	1113	127
·993	·745 212	6939	1414	187
·994	·752 151	7755	1865	294
0·995	0·759 906			

TABLE 5

u	sm 2u	δ¹	2δ²
0·75	0·993 701	+	
		1310	-405
·76	·995 011	1116	-374
·77	·996 127	936	-344
·78	·997 063	772	-312
·79	·997 835	624	-282
0·80	0·998 459		
		490	-250
·81	·998 949	374	-217
·82	·999 323	273	-187
·83	·999 596	187	-155
·84	·999 783	118	-122
0·85	0·999 901		
		65	-90
·86	·999 966	28	-59
·87	0·999 994	6	-27
·88	1·000 000	1	+5
·89	1·000 001	11	+38
0·90	1·000 012		
		39	+70
·91	1·000 051	81	+100
·92	1·000 132	139	+134
·93	1·000 271	215	+166
·94	1·000 486	305	+197
0·95	1·000 791		
		412	+230
·96	1·001 203	535	+263
·97	1·001 738	675	+295
·98	1·002 413	830	+327
0·99	1·003 243	1002	+361
1·00	1·004 245		
		1191	+394
1·01	1·005 436	1396	+428
1·02	1·006 832	1619	+463
1·03	1·008 451	1859	+497
1·04	1·010 310	2116	+531
1·05	1·012 426		

For values of sm 2u greater than 1·005, evaluate the reciprocal. Enter Table 4 with this reciprocal as argument, and subtract the derived value of u from 1·766 639.

For example, if sm 2u = 1·024 763, the reciprocal is 0·975 835. With this value as argument, we find the table gives u = 0·673 990. The required value is therefore u = 1·092 649.

TABLES FOR FINDING u FROM sm $2u$

TABLE 6

$-$sm $2u$	$-u$	δ^1	$2\delta^2$	$-$sm $2u$	$-u$	δ^1	$2\delta^2$
0·00	0·000 000	+	−	0·50	0·245 080	+	−
		5000				4612	43
·01	·005 000	5000		·51	·249 692	4590	42
·02	·010 000	5000		·52	·254 282	4570	43
·03	·015 000	5000		·53	·258 852	4547	46
·04	·020 000	4999		·54	·263 399	4524	47
0·05	0·024 999			0·55	0·267 923		
		5000				4500	47
·06	·029 999	4999		·56	·272 423	4477	48
·07	·034 998	4999		·57	·276 900	4452	50
·08	·039 997	4998		·58	·281 352	4427	50
·09	·044 995	4997		·59	·285 779	4402	51
0·10	0·049 992			0·60	0·290 181		
		4996				4376	53
·11	·054 988	4995		·61	·294 557	4349	53
·12	·059 983	4993		·62	·298 906	4323	54
·13	·064 976	4992		·63	·303 229	4295	56
·14	·069 968	4990		·64	·307 524	4267	56
0·15	0·074 958			0·65	0·311 791		
		4987				4239	57
·16	·079 945	4986		·66	·316 030	4210	57
·17	·084 931	4982		·67	·320 240	4182	58
·18	·089 913	4979	7	·68	·324 422	4152	60
·19	·094 892	4975	7	·69	·328 574	4122	59
0·20	0·099 867			0·70	0·332 696		
		4972	8			4093	60
·21	·104 839	4967	10	·71	·336 789	4062	61
·22	·109 806	4962	10	·72	·340 851	4032	61
·23	·114 768	4957	10	·73	·344 883	4001	62
·24	·119 725	4952	12	·74	·348 884	3970	63
0·25	0·124 677			0·75	0·352 854		
		4945	13			3938	63
·26	·129 622	4939	13	·76	·356 792	3907	62
·27	·134 561	4932	15	·77	·360 699	3876	64
·28	·139 493	4924	15	·78	·364 575	3843	64
·29	·144 417	4917	17	·79	·368 418	3812	64
0·30	0·149 334			0·80	0·372 230		
		4907	18			3779	65
·31	·154 241	4899	19	·81	·376 009	3747	64
·32	·159 140	4888	20	·82	·379 756	3715	64
·33	·164 028	4879	21	·83	·383 471	3683	66
·34	·168 907	4867	22	·84	·387 154	3649	65
0·35	0·173 774			0·85	0·390 803		
		4857	23			3618	64
·36	·178 631	4844	26	·86	·394 421	3585	66
·37	·183 475	4831	25	·87	·398 006	3552	65
·38	·188 306	4819	27	·88	·401 558	3520	65
·39	·193 125	4804	29	·89	·405 078	3487	66
0·40	0·197 929			0·90	0·408 565		
		4790	29			3454	64
·41	·202 719	4775	30	·91	·412 019	3423	65
·42	·207 494	4760	32	·92	·415 442	3389	65
·43	·212 254	4743	34	·93	·418 831	3358	64
·44	·216 997	4726	34	·94	·422 189	3325	65
0·45	0·221 723			0·95	0·425 514		
		4709	35			3293	64
·46	·226 432	4691	38	·96	·428 807	3261	64
·47	·231 123	4671	38	·97	·432 068	3229	64
·48	·235 794	4653	38	·98	·435 297	3197	63
·49	·240 447	4633	41	0·99	·438 494	3166	63
0·50	0·245 080			1·00	0·441 660		

For values of sm $2u$ less than −1, evaluate the reciprocal. Enter Table 6 with this reciprocal as argument, and subtract the derived of u from −0·883 319. For example, if sm $2u$ = −2.231 610, the reciprocal is −0·448 107. With this value as argument, we find the table gives u = −0·220 830. The required value is therefore u = −0·662 489.

TABLES FOR FINDING v FROM $sm^3(2v/\sqrt{3})$

TABLE 7

v	$sm^3(2v/\sqrt{3})$	δ^1	$2\delta^2$
0·00	0·000 000	+	+
		2	12
·01	·000 002		
		10	28
·02	·000 012		
		30	47
·03	·000 042		
		57	63
·04	·000 099		
		93	83
0·05	0·000 192		
		140	103
·06	·000 332		
		196	120
·07	·000 528		
		260	138
·08	·000 788		
		334	156
·09	·001 122		
		416	175
0·10	0·001 538		
		509	194
·11	·002 047		
		610	211
·12	·002 657		
		720	229
·13	·003 377		
		839	247
·14	·004 216		
		967	264
0·15	0·005 183		
		1103	283
·16	·006 286		
		1250	300
·17	·007 536		
		1403	316
·18	·008 939		
		1566	333
·19	·010 505		
		1736	350
0·20	0·012 241		
		1916	367
·21	·014 157		
		2103	382
·22	·016 260		
		2293	398
·23	·018 558		
		2501	412
·24	·021 059		
		2710	427
0·25	0·023 769		
		2928	443
·26	·026 697		
		3153	455
·27	·029 850		
		3383	468
·28	·033 233		
		3621	481
·29	·036 854		
		3864	492
0·30	0·040 718		
		4113	505
·31	·044 831		
		4369	515
·32	·049 200		
		4628	523
·33	·053 828		
		4892	534
·34	·058 720		
		5162	543
0·35	0·063 882		

TABLE 8

$-v$	$-sm^3(2v/\sqrt{3})$	δ^1	$2\delta^2$
0·00	0·000 000	+	+
		2	12
·01	·000 002		
		10	28
·02	·000 012		
		30	47
·03	·000 042		
		57	63
·04	·000 099		
		93	84
0·05	0·000 192		
		141	102
·06	·000 333		
		195	120
·07	·000 528		
		261	139
·08	·000 789		
		334	157
·09	·001 123		
		418	176
0·10	0·001 541		
		510	195
·11	·002 051		
		613	214
·12	·002 664		
		724	233
·13	·003 388		
		846	252
·14	·004 234		
		976	270
0·15	0·005 210		
		1116	291
·16	·006 326		
		1267	310
·17	·007 593		
		1426	330
·18	·009 019		
		1597	351
·19	·010 616		
		1777	370
0·20	0·012 393		
		1967	392
·21	·014 360		
		2169	413
·22	·016 529		
		2380	434
·23	·018 909		
		2603	456
·24	·021 512		
		2836	479
0·25	0·024 348		
		3082	502
·26	·027 430		
		3338	525
·27	·030 768		
		3607	551
·28	·034 375		
		3889	575
·29	·038 264		
		4182	601
0·30	0·042 446		
		4490	628
·31	·046 936		
		4810	654
·32	·051 746		
		5144	684
·33	·056 890		
		5494	713
·34	·062 384		
		5857	743
0·35	0·068 241		

For values of $sm^3(2v/\sqrt{3})$ from 0·95 to 1·00, evaluate the complement $1 - sm^3(2v/\sqrt{3})$. Enter Table 7 with this complement as respondent, and subtract the derived value of v from 1·529 954.

TABLES FOR FINDING v FROM $sm^3(2v/\sqrt{3})$

TABLE 9

$sm^3(2v/\sqrt{3})$	v	δ^1	$2M^2$	δ^3	$sm^3(2v/\sqrt{3})$	v	δ^1	$2M^2$	δ^3
0·05	0·321 771	+	−	+	0·50	0·764 977	+	+	+
		20758	4731	759			7275	3	
·06	·342 529	18694	3567	481	·51	·772 252	7278	12	
·07	·361 223	17111	2795	325	·52	·779 530	7287	20	
·08	·378 334	15853	2257	231	·53	·786 817	7298	27	
·09	·394 187	14826	1862	173	·54	·794 115	7314	36	
0·10	0·409 013				0·55	0·801 429			
		13972	1566	129			7334	43	
·11	·422 985	13247	1339	103	·56	·808 763	7357	52	
·12	·436 232	12625	1154	83	·57	·816 120	7386	61	
·13	·448 857	12086	1007	66	·58	·823 506	7418	68	
·14	·460 943	11613	887	55	·59	·830 924	7454	79	
0·15	0·472 556				0·60	0·838 378			
		11195	787				7497	89	
·16	·483 751	10823	701		·61	·845 875	7543	97	
·17	·494 574	10491	630		·62	·853 418	7594	108	
·18	·505 065	10191	567		·63	·861 012	7651	119	
·19	·515 256	9922	514		·64	·868 663	7713	131	
0·20	0·525 178				0·65	0·876 376			
		9676	468				7782	142	
·21	·534 854	9453	425		·66	·884 158	7855	154	
·22	·544 307	9250	389		·67	·892 013	7936	170	
·23	·553 557	9063	358		·68	·899 949	8025	184	
·24	·562 620	8892	327		·69	·907 974	8120	199	
0·25	0·571 512				0·70	0·916 094			
		8736	300				8224	217	
·26	·580 248	8592	277		·71	·924 318	8337	235	
·27	·588 840	8459	255		·72	·932 655	8459	255	
·28	·597 299	8337	235		·73	·941 114	8592	277	
·29	·605 636	8224	217		·74	·949 706	8736	300	
0·30	0·613 860				0·75	0·958 442			
		8120	199				8892	327	
·31	·621 980	8025	184		·76	·967 334	9063	358	
·32	·630 005	7936	170		·77	·976 397	9250	380	
·33	·637 941	7855	154		·78	·985 647	9453	425	
·34	·645 796	7732	142		·79	·995 100	9676	468	
0·35	0·653 578				0·80	1·004 776			
		7713	131				9922	514	
·36	·661 291	7651	119		·81	1·014 698	10191	567	
·37	·668 942	7594	108		·82	1·024 889	10491	630	
·38	·676 536	7543	97		·83	1·035 380	10823	701	
·39	·684 079	7497	89		·84	1·046 203	11195	787	
0·40	0·691 576				0·85	1·057 398			
		7454	79				11613	887	55
·41	·699 030	7418	68		·86	1·069 011	12086	1007	66
·42	·706 448	7386	61		·87	1·081 097	12625	1154	83
·43	·713 834	7357	52		·88	1·093 722	13247	1339	103
·44	·721 191	7334	43		·89	1·106 969	13972	1566	129
0·45	0·728 525				0·90	1·120 941			
		7314	36				14826	1862	173
·46	·735 839	7298	27		·91	1·135 767	15853	2257	231
·47	·743 137	7287	20		·92	1·151 620	17111	2795	325
·48	·750 424	7278	12		·93	1·168 731	18694	3567	481
·49	·757 702	7275	3		·94	1·187 425	20758	4731	759
0·50	0·764 977				0·95	1·208 183			

TABLES FOR FINDING v FROM $sm^3(2v/\sqrt{3})$

TABLE 10

$-sm^3(2v/\sqrt{3})$	$-v$	δ^1	$2M^2$	δ^3	$-sm^3(2v/\sqrt{3})$	$-v$	δ^1	$2\delta^2$
0·00	0·000 000	+	−	+	0·50	0·640 569	+	−
·01	·186 270				·51	·644 035	3466	123
·02	·234 299	See Table 8			·52	·647 441	3406	118
·03	·267 768				·53	·650 789	3348	114
·04	·294 239				·54	·654 081	3292	111
							3237	107
0·05	0·316 449				0·55	0·657 318		
·06	·335 741	19292	4911	768	·56	·660 503	3185	103
·07	·352 883	17142	3731	489	·57	·663 637	3134	101
·08	·368 364	15481	2943	332	·58	·666 721	3084	98
·09	·382 516	14152	2394	235	·59	·669 757	3036	95
		13058	1992	177			2989	92
0·10	0·395 574				0·60	0·672 746		
·11	·407 715	12141	1688	133	·61	·675 690	2944	89
·12	·419 072	11357	1454	105	·62	·678 590	2900	87
·13	·429 750	10678	1265	86	·63	·681 447	2857	84
·14	·439 835	10085	1115	66	·64	·684 263	2816	83
		9558	992	59			2774	80
0·15	0·449 393				0·65	0·687 037		
·16	·458 483	9090	886		·66	·689 773	2736	77
·17	·467 152	8669	800		·67	·692 470	2697	76
·18	·475 440	8288	726		·68	·695 130	2660	73
·19	·483 381	7941	663		·69	·697 754	2624	72
		7624	605				2588	70
0·20	0·491 005				0·70	0·700 342		
·21	·498 339	7334	558		·71	·702 896	2554	69
·22	·505 404	7065	518		·72	·705 415	2519	67
·23	·512 220	6816	479		·73	·707 902	2487	64
·24	·518 806	6586	444		·74	·710 357	2455	63
		6372	416				2424	62
0·25	0·525 178				0·75	0·712 781		
·26	·531 348	6170	389		·76	·715 174	2393	61
·27	·537 331	5983	364		·77	·717 537	2363	59
·28	·543 137	5806	343		·78	·719 871	2334	58
·29	·548 777	5640	323		·79	·722 176	2305	56
		5433	304				2278	55
0·30	0·554 260				0·80	0·724 454		
·31	·559 596	5336	288		·81	·726 704	2250	55
·32	·564 791	5195	274		·82	·728 927	2223	52
·33	·569 853	5062	258		·83	·731 125	2198	51
·34	·574 790	4937	246		·84	·733 297	2172	51
		4816	234				2147	49
0·35	0·579 606				0·85	0·735 444		
·36	·584 309	4703	223		·86	·737 567	2123	49
·37	·588 902	4593	213		·87	·739 665	2098	47
·38	·593 392	4490	203		·88	·741 741	2076	46
·39	·597 782	4390	195		·89	·743 793	2052	46
		4295	186				2030	45
0·40	0·602 077				0·90	0·745 823		
·41	·606 281	4204	178		·91	·747 830	2007	43
·42	·610 398	4117	171		·92	·749 817	1987	42
·43	·614 431	4033	165		·93	·751 782	1965	43
·44	·618 383	3952	158		·94	·753 726	1944	41
		3875	151				1924	40
0·45	0·622 258				0·95	0·755 650		
·46	·626 059	3801	146		·96	·757 554	1904	40
·47	·629 788	3729	142		·97	·759 438	1884	39
·48	·633 447	3659	136		·98	·761 303	1865	38
·49	·637 040	3593	130		0·99	·763 149	1846	37
		3529	127				1828	36
0·50	0·640 569				1·00	0·764 977		

For values of $sm^3(2v/\sqrt{3})$ less than −1, evaluate the reciprocal. Enter Table 10 with this reciprocal as argument, and subtract the derived value of v from −1·529 954.

TABLES FOR FINDING u FROM cn $2\sqrt{2}u$ ($k = k' = \sin 45°$)

TABLE 11

cn $2\sqrt{2}u$	u	δ^1	$2\delta^2$	cn $2\sqrt{2}u$	u	δ^1	$2u^2$	δ^3
0·00	0·655 514	–	–	0·50	0·403 910	–	–	–
·01	·650 514	5000		·51	·398 739	5171	29	
·02	·645 514	5000		·52	·393 553	5186	31	
·03	·640 514	5000		·53	·388 351	5202	32	
·04	·635 514	5000		·54	·383 133	5218	34	
0·05	0·630 514	5000		0·55	0·377 897	5236	38	
·06	·625 514	5000		·56	·372 641	5256	40	
·07	·620 514	5000		·57	·367 365	5276	42	
·08	·615 514	5000		·58	·362 067	5298	45	
·09	·610 514	5000		·59	·356 746	5321	49	
0·10	0·605 514	5000		0·60	0·351 399	5347	52	
·11	·600 514	5001		·61	·346 026	5373	54	
·12	·595 513	5000		·62	·340 625	5401	59	
·13	·590 513	5001		·63	·335 193	5432	63	
·14	·585 512	5001		·64	·329 729	5464	66	
0·15	0·580 511	5002		0·65	0·324 231	5498	72	
·16	·575 509	5002		·66	·318 695	5536	76	
·17	·570 507	5002		·67	·313 121	5574	81	
·18	·565 505	5003		·68	·307 504	5617	89	
·19	·560 502	5004		·69	·301 841	5663	93	
0·20	0·555 498	5004		0·70	0·296 131	5710	99	
·21	·550 494	5005		·71	·290 369	5762	108	
·22	·545 489	5007		·72	·284 551	5818	116	
·23	·540 482	5007		·73	·278 673	5878	124	
·24	·535 475	5010		·74	·272 731	5942	133	
0·25	0·530 465	5010		0·75	0·266 720	6011	143	
·26	·525 455	5013		·76	·260 635	6085	156	
·27	·520 442	5014		·77	·254 468	6167	169	
·28	·515 428	5016		·78	·248 214	6254	182	
·29	·510 412	5020		·79	·241 865	6349	198	
0·30	0·505 392	5021		0·80	0·235 413	6452	217	
·31	·500 371	5025		·81	·228 847	6566	237	
·32	·495 346	5028	7	·82	·222 158	6689	259	
·33	·490 318	5032	8	·83	·215 333	6825	287	
·34	·485 286	5036	8	·84	·208 357	6976	317	
0·35	0·480 250	5040	9	0·85	0·201 215	7142	353	
·36	·475 210	5045	10	·86	·193 886	7329	395	
·37	·470 165	5050	11	·87	·186 349	7537	445	
·38	·465 115	5056	12	·88	·178 575	7774	507	
·39	·460 059	5062	13	·89	·170 531	8044	530	
0·40	0·454 997	5069	14	0·90	0·162 177	8354	674	54
·41	·449 928	5076	14	·91	·153 459	8718	785	64
·42	·444 852	5083	16	·92	·144 313	9146	936	91
·43	·439 769	5092	18	·93	·134 648	9665	1144	122
·44	·434 677	5101	19	·94	·124 342	10306	1430	175
0·45	0·429 576	5111	20	0·95	0·113 220	11122	1851	271
·46	·424 465	5121	22	·96	·101 011			
·47	·419 344	5133	23	·97	·087 257			
·48	·414 211	5144	24	·98	·071 066	See Table 13		
·49	·409 067	5157	27	0·99	·050 125			
0·50	0·403 910			1·00	0·000 000			

TABLES FOR FINDING u FROM $\operatorname{cn} 2\sqrt{2}u$ ($k = k' = \sin 45°$)

TABLE 12

$-\operatorname{cn} 2\sqrt{2}u$	u	δ^1 +	$2\delta^2$ +	$-\operatorname{cn} 2\sqrt{2}u$	u	δ^1 +	$2M^2$ +	δ^3 +
0·00	0·655 514	5000		0·50	0·907 119	5171	29	
·01	·660 514	5000		·51	·912 290	5186	30	
·02	·665 514	5000		·52	·917 476	5201	33	
·03	·670 514	5000		·53	·922 677	5219	35	
·04	·675 514	5000		·54	·927 896	5236	37	
0·05	0·680 514	5000		0·55	0·933 132	5256	40	
·06	·685 514	5000		·56	·938 388	5276	42	
·07	·690 514	5001		·57	·943 664	5298	45	
·08	·695 515	5000		·58	·948 962	5321	49	
·09	·700 515	5000		·59	·954 283	5347	51	
0·10	0·705 515	5000		0·60	0·959 630	5372	55	
·11	·710 515	5001		·61	·965 002	5402	59	
·12	·715 516	5000		·62	·970 404	5431	62	
·13	·720 516	5001		·63	·975 835	5464	68	
·14	·725 517	5001		·64	·981 299	5499	71	
0·15	0·730 518	5002		0·65	0·986 798	5535	76	
·16	·735 520	5001		·66	·992 333	5575	82	
·17	·740 521	5003		·67	0·997 908	5617	87	
·18	·745 524	5003		·68	1·003 525	5662	94	
·19	·750 527	5003		·69	1·009 187	5711	100	
0·20	0·755 530	5005		0·70	1·014 898	5762	107	
·21	·760 535	5005		·71	1·020 660	5818	116	
·22	·765 540	5007		·72	1·026 478	5878	123	
·23	·770 547	5007		·73	1·032 356	5941	133	
·24	·775 554	5009		·74	1·038 297	6011	145	
0·25	0·780 563	5011		0·75	1·044 308	6086	156	
·26	·785 574	5012		·76	1·050 394	6167	168	
·27	·790 586	5015		·77	1·056 561	6254	182	
·28	·795 601	5016		·78	1·062 815	6349	198	
·29	·800 617	5019		·79	1·069 164	6452	216	
0·30	0·805 636	5022		0·80	1·075 616	6565	238	
·31	·810 658	5025		·81	1·082 181	6690	260	
·32	·815 683	5028	7	·82	1·088 871	6825	286	
·33	·820 711	5032	8	·83	1·095 696	6976	317	
·34	·825 743	5036	8	·84	1·102 672	7142	353	
0·35	0·830 779	5040	9	0·85	1·109 814	7329	395	
·36	·835 819	5045	10	·86	1·117 143	7537	445	
·37	·840 864	5050	11	·87	1·124 680	7774	507	
·38	·845 914	5056	12	·88	1·132 454	8044	580	
·39	·850 970	5062	13	·89	1·140 498	8354	673	53
0·40	0·856 032	5069	13	0·90	1·148 852	8717	787	67
·41	·861 101	5075	15	·91	1·157 569	9147	938	88
·42	·866 176	5084	17	·92	1·166 716	9665	1143	123
·43	·871 260	5092	17	·93	1·176 381	10306	1429	175
·44	·876 352	5101	19	·94	1·186 687	11122	1851	271
0·45	0·881 453	5111	20	0·95	1·197 809			
·46	·886 564	5121	21	·96	1·210 018			
·47	·891 685	5132	24	·97	1·223 771			
·48	·896 817	5145	25	·98	1·239 963	See Table 14		
·49	·901 962	5157	26	0·99	1·260 903			
0·50	0·907 119			1·00	1·311 029			

TABLES FOR FINDING u FROM $cn\ 2\sqrt{2}u$ $(k = k' = \sin 45°)$

	TABLE 13					TABLE 14		
u	$cn\ 2\sqrt{2}u$	δ^1	$2\delta^2$		u	$-cn\ 2\sqrt{2}u$	δ^1	$2\delta^2$
0·00	1·000 000	–	–		1·15	0·901 342	+	–
·01	0·999 600	400	1599		1·16	·912 707	11365	1193
·02	·998 401	1199	1595		1·17	·923 464	10757	1239
·03	·996 406	1995	1587		1·18	·935 590	10126	1281
·04	·993 620	2786	1575		1·19	·943 066	9476	1321
		3570	1561				8305	1361
0·05	0·990 050				1·20	0·951 871		
·06	·985 703	4347	1543		1·21	·959 986	8115	1397
·07	·980 590	5113	1520		1·22	·967 394	7408	1432
·08	·974 723	5867	1495		1·23	·974 077	6683	1464
·09	·968 115	6608	1467		1·24	·980 021	5944	1492
		7334	1435				5191	1518
0·10	0·960 781				1·25	0·985 212		
·11	·952 738	8043	1401		1·26	·989 638	4426	1540
·12	0·944 003	8735	1364		1·27	·993 289	3651	1559
					1·28	·996 156	2867	1574
					1·29	·998 233	2077	1586
							1281	1595
					1·30	0·999 514		
					1·31	·999 996	482	1599
					1·32	0·999 678	–318	1599

General Information

Each monograph in this series will cover one aspect of cartography and will comprise either a single major work, or a collection of research papers relevant to a principal theme.

Subjects to be chosen for the monographs will include the many topics that comprise cartography, e.g., the history and development of cartography, topographic and thematic mapping, analysis of national and regional atlases, the techniques of map production, automation in cartography, map design, map projections and map librarianship.

This series of monographs complements *The Canadian Cartographer.*

Subscriptions

Subscriptions may be entered on an annual or standing order basis for *Cartographica.* The subscription rate is $12.00 for the three monographs. Single copies are $4.00 each. Orders may be placed direct with the publisher or through any agency. The monographs may be ordered as supplements to *Canadian Cartographer* at a combined subscription rate of $15.00 for 5 issues (2 numbers of *Canadian Cartographer* and 3 monographs per year).

Address enquiries to:

General Editor, Department of Geography, York University
4700 Keele Street, Toronto, Canada
Executive Secretary: Yvonne Wood, Geography, York University, Toronto

A 4

Monographs

1971 No. 1. The 17th Century Cartography of Newfoundland. By Fabian O'Dea. 48 p.

No. 2. Map Design and the Map User. Papers selected from the Symposium on the Influence of the Map User on Map Design, held at Queen's University, Kingston, Canada, September 1970. 84 p.

No. 3. Russian Maps and Atlases as Historical Sources. By Leonid A. Goldenberg. 76 p.

1972 No. 4. National Atlases. K.A. Salichtchev, Ed. 81 p.

No. 5. Eskimo Maps from the Canadian Eastern Arctic. By John Spink and D.W. Moodie, 100 p.

No. 6. Explorers' Maps of the Canadian Arctic 1818-1860. By Coolic Verner. 84 p. (maps 12 p.)

1973 No. 7. Economic Maps in Pre-Reform Russia. By Arkady I. Preobrazhensky. 46 p.

No. 8. Land Surveys of Southern Ontario. An introduction and index to the field notebooks of the Ontario Land Surveyors, 1784-1859. By Louis Gentilcore and Kate Donkin. 116 p.

No. 9. Computer Cartography in Canada. A collection of papers coordinated and edited by Aubrey L. LeBlanc. 103 p.

1974 No. 10. Cartographic Generalisation. Some Concepts and Explanation. By H.J. Steward. 77 p.

No. 11. The Seven Aspects of a General Map Projection. By Thomas Wray. 72 p.

No. 12. Index and Abstracts, Canadian Cartographer Vols. 1-10 1964-1973. 164 p.

1975 No. 13. Essays on the History of Russian Cartography 16th to 19th Centuries. 145 p.

No. 14. Studies in the History of Russian Cartography. By Fyodor A. Shibanov. Part 1. 101 p.

No. 15. Studies in the History of Russian Cartography. By Fyodor A. Shibanov. Part 2. 86 p.

DESIGNED BY

HAROLD KURSCHENSKA

COMPOSED BY L.P. LEE

PRINTED ON ROLLAND TINT, ASH GREY BY

THE UNIVERSITY OF TORONTO PRESS